AF553875

A-Z
NANOPHYSICS

A-Z
NANOPHYSICS

Prof. Girish Upadhayay

CENTRUM PRESS
NEW DELHI-110002 (INDIA)

CENTRUM PRESS
H.O.: 4360/4, Ansari Road, Daryaganj,
New Delhi-110 002 (India)
Ph.: 23278000, 23261597

B.O.: No. 1015, Ist Main Road, BSK IIIrd Stage
IIIrd Phase, IIIrd Block,
Bangalore - 560 085 (India)
Tel.: 080-41723429
Visit us at: www.centrumpress.com

A-Z Nanophysics

First Edition, 2009

ISBN 978-93-80252-00-1

PRINTED IN INDIA

Printed at Mehra Offset Press, Delhi

Contents

Preface

"This book has been brought out to acquaint the readers with the hype and the rigmarole behind Nano Physics. The emphasis of the book is on presenting forth a thorough introductory manual on the subject, detailing its history, the science and the practice, the techniques, the trends and developments, of this science.

The multifold implications of nano physics have been discussed most comprehensively, as this science holds the key to understanding the future of humanity in terms of scientific developments. The critical insights presented forth in the book are highly informative, and it serves well for all those interested in understanding Nano Physics in all its machinations, and manifestations."

Author

Chapter 1

Nanoscience

Nanoscience is the study of phenomena and manipulation of materials at atomic, molecular and macromolecular scales, where properties differ significantly from those at larger scales. Eight to ten atoms span one nanometer (nm). A human hair is approximately 80,000 nm thick. Typically nanoscience is concerned with the study of objects ranging in size from about 100 nanometres downwards.

Nanotechnology is the design, characterisation, production and application of structures, devices and systems by controlling shape and size at the nanoscale. By creating nanometer-scale structures it is possible to control fundamental characteristics of a material including its melting point, magnetic properties, colour, without changing the material's chemical composition.

The goal of nanotechnology is to direct atoms and molecules to form desired structures or patterns with novel functionality. At the nanoscale, the physical, chemical, and biological properties of materials differ in fundamental and valuable ways from the properties of individual atoms and molecules or bulk matter. Nanotechnology R&D is directed toward understanding and creating improved materials, devices, and systems that exploit these new properties.

Although nanotechnology is a relatively new research area nanotechnologies are already emerging that will change our lives in unforeseeable ways in the coming decades, and the range of possible future applications is constantly growing. Nanoscale materials are used in electronic, magnetic and optoelectronic, biomedical, pharmaceutical, cosmetic, energy, catalytic and materials applications. Areas producing the greatest revenue for nanoparticles are chemical-mechanical polishing, magnetic recording tapes, sunscreens, vehicular catalyst supports, biolabeling, electroconductive coatings and optical fibers. Today most computer hard drives contain giant magnetoresistance (GMR) heads that, through nanothin layers of magnetic materials, allow for an order of magnitude increase in storage capacity. Other electronic applications include non-volatile magnetic memory, vehicular sensors, landmine detectors and solid-state compasses.

APPLICATIONS AND POTENTIAL BENEFITS OF NANOTECHNOLOGY

Using nanotechnology materials with distinct properties can be fabricated. Nanotechnologically improved products rely on a change in the physical properties when the feature sizes are shrunk. Nanoparticles for example take advantage of their dramatically increased surface area to volume ratio. Their optical properties, e.g. fluorescence, become a function of the particle diameter. When brought into a bulk material, nanoparticles can strongly influence the mechanical properties of the material such as the stiffness or elasticity. Traditional polymers can be reinforced by nanoparticles resulting in novel materials e.g. as lightweight replacements for metals. Such nanotechnologically enhanced materials will provide a weight reduction accompanied by an increase in stability and an improved mechanical functionality. An increasing benefit to society is anticipated from the development of these materials.

Nanoscale devices and nanoscale components of larger

devices are of the same size as biological entities. They are smaller than human cells (10,000 to 20,000 nanometers in diameter) and similar in size to large biological macromolecules such as enzymes. Nanoscale devices smaller than 50 nanometers can enter most cells easily. As a result nanoscale devices can readily interact with biomolecules on both the cell surface and within the cell, often in ways that do not alter the behaviour and biochemical properties of those molecules. Nanoscale devices are already proving that they can deliver therapeutic agents that can act where they are likely to be most effective i.e. within the cell. Nanoparticles and nanoprobe techniques are currently being used to identify where plaque is building to dangerous levels in arteries. Fullerenes1 are being developed to carry drugs and radioactive molecules to blood vessels that supply cancer cells. The actual construction and characterisation of nanoscale devices may contribute to understanding the development of cancerous cells.

Magnetic nanoparticles offer an effective and reliable method to remove heavy metal contaminants from waste water by making use of magnetic separation techniques. Using nanoscale particles increases the efficiency to absorb contaminants and is comparatively inexpensive compared to traditional precipitation and filtration methods

The most advanced nanotechnology projects related to energy are: storage, conversion, manufacturing improvements by reducing materials and process rates, energy saving e.g. by better thermal insulation, and enhanced renewable energy sources applications. Typical applications can be found in light bulbs, solar cells and fuel cells to name a few.

The production of displays with low energy consumption could be accomplished using carbon nanotubes 2(CNT). CNTs can be electrically conductive and due to their small diameter of several nanometers they can be used as field emitters with extremely high efficiency for field emission displays. The principle of operation resembles that of the cathode ray tube, but on a much smaller length scale.

Nanotechnology is already impacting the field of consumer goods, providing products with novel functions ranging from easy-to-clean to scratch-resistant. Wrinkle-resistant and stain-repellent textiles based on nanoparticles impregnation are already available. In the field of cosmetics, novel products utilising nanotechnology offers one of the most lucrative applications of nanoscience.

CURRENT RESEARCH IN NANOTECHNOLOGIES

The Centre for Research on Adaptive Nanostructures and Nanodevices (CRANN), Trinity College Dublin (TCD), is an internationally recognised centre of excellence in nanoscience. It is a dedicated nanoscience research centre with 80 researchers working on the latest nanoscience research with a mission to advance the frontiers of nanoscience. CRANN is comprised of partners from academia, TCD and University College Cork (UCC), along with industry partners including Intel, Hewlett-Packard, and a number of Irish high-technology companies.

The researchers are housed in Ireland's first purpose-built research institute. The building has have world-class facilities to house the activities of over 120 scientists, with ultra-low vibration laboratories to allow highly sensitive measurements of nanoscale structures, and state-of-the-art clean rooms where even particles of dusts are carefully filtered out to allow high-purity fabrication of these tiny objects. CRANN is set to put Ireland on the map as a global player in this revolutionary new field by combining its scientific excellence with the support of SFI and TCD, the expertise of Intel Ireland, Hewlett-Packard Manufacturing Ltd. and our other industrial partners, and the private philanthropic interest of Dr. Martin Naughton.

One of CRANN's main interest areas is in manipulation of spin i.e. replacing electrons as information carrying agent in electronic devices. There is a limitation on how small we can make chips. Moving electrons around a chip costs a lot, both in terms of energy and energy dissipation. Conventional electronics ignore the spin of the electron. The research groups of two of our Principal Investigators are looking at the dynamics of spin, and investigating novel structures to test spin based devices. Spin-polarised electron transport in nanoscale devices and spin electronic properties of magnetic oxides are under investigation. These materials have a lot of potential for the future developments of information technology such as novel electronic devices including spin logic and information storage cells. The new science of spin electronics will deliver smaller, faster devices with novel properties with reduced energy dissipation.

Miniaturisation of devices to the nanoscale is of current interest, specifically the chemistry and materials properties of nanoscale systems that have potential applications in the areas of devices and sensors. We are interested in the challenge of processing traditional device materials such as silicon so as to enable the successful fabrication of the next generation computer chips. To determine the limits of this technology and investigate the possibility of tweaking these material systems to get faster devices it is necessary to study the detailed nature of these processes at an atomic level. Much of our research involves the use of scanning tunnelling microscopy and atomic force microscopy to characterise both the materials themselves and their electrical and mechanical performance.

Intel and CRANN researchers are seeking to identify new approaches to fabricating future generations of computer chips using nanotechnology. This research collaboration has led to the development of the Adaptive Grid Substrate (AGS), which facilitates testing materials and devices at the nanometre scale. The AGS addresses the issue of how to arrange objects at the nanoscale. Silicon chips have been patterned and functionalised to act as templates for the arrangement of materials such as carbon nanotubes.

The trick is to encourage the nanotubes to bed down in predefined trenches on the silicon grid pattern. This allows researchers to find them, measure them and gain much more reliable information about these materials and their distinct properties. The research is motivated by the need to make increasing smaller devices, beyond that currently possible using CMOS technology. The challenge facing researchers in this field is the difficulty in characterising large numbers of potential nanoscale device components in a manner that is compatible with chip manufacturing technologies. The AGS provides direct electrical connections to these nanocomponents for electrical testing and also specially designed relief features that promote ordering of these components over the entire substrate. Ultimately it is hoped that the AGS will be developed to allow direct links between CMOS and the nanoworld, thereby extending current technologies and manufacturing processes while at the same time benefiting from the increased device integration densities afforded by nanotechnology.

It is estimated that there are 20,000 researchers working in nanotechnology worldwide today and that in excess of $2 billion in worldwide government funding was spent on nanotechnology R&D in 2002. In the US over $1 billion of federal funding was spent on nanotechnology R&D in 2005. With a current estimated value of $20 billion, nanotechnology is projected to grow to a $1 trillion global industry by 2015, and 2 million workers will be needed to support those industries.

Chapter 2

Advances in Nanomagnetism Via X-Ray Techniques

The present paper is a distillation of a workshop, entitled "Nanomagnetism Using X-ray Techniques" that was held in Fontana, Wisconsin, August 29-September 1, 2004, that was held as part of a strategic planning exercise sponsored by the Advanced Photon Source (APS) of the Argonne National Laboratory. The APS is a third-generation hard X-ray synchrotron source that serves as a major national user facility and is supported by the US Department of Energy. The purpose of the workshop was two-fold: first, to survey topics of current interest in nanomagnetism and identify emerging major themes. Second, to evaluate the role that X-ray techniques could play in addressing the fundamental research issues that lie ahead. Enhancing present X-ray capabilities and/ or evolving new techniques to meet future challenges will assist both the synchrotron

X-ray science community in general and the APS in particular in planning for new instrumentation and resource priorities. The workshop was organized along three broad thematic areas in nanomaterials realm: confined magnetism, cluster magnetism, and complex oxide/phase-separated magnetic systems. We here first present an overview of grand challenges in nanomagnetics and then summarize the key basic and applied aspects of new nanostructured magnetic materials. The role that synchrotron X-ray techniques play in addressing the grand challenges follows. Finally, the paper ends with recommendations for future directions.

Nanomagnetism describes the science and technology underlying the magnetic behaviour of nanostructured (1–100 nm) systems. It focuses on the magnetic behaviour of individual building blocks of nanostructured systems as well as on combinations of individual building blocks that display collective magnetic phenomena. A fundamental understanding of nanomagnetism can be exploited to yield integrated systems with complex structures and architectures that possess new functionalities. Nanomagnetism provides new paradigms for condensed matter science.

For example, there are differences in the temperature stability of a ferromagnetic state, spin polarization effects that can induce magnetic moments on nearby "non-magnetic" atoms, spin effects on quantum mechanical tunneling probabilities, the possibility of directing the movement of nanomagnets within individual living cells, the potential to detect pathogens through a combination of biological molecular recognition and magnetoresistive sensing using nanomagnetic thin films. Fundamental scientific questions common to the potential technological applications envisioned are concerned with the origin of magnetic couplings, spin transport across interfaces, spin-lattice interactions in complex materials and the magnetic domain configuration and dynamics arising from contact between different kind of magnets, for example antiferromagnets and ferromagnets.

GRAND CHALLENGES IN NANOMAGNETISM

The grand challenges in nanomagnetism can be expressed in a number of ways depending on the audience. At the most basic level the challenge is to create, explore and understand new materials that exhibit unexpected collective behaviour, also known as emergent behaviour. The creation of emergent matter energizes the synthesis community, and high-quality synthesis of new materials is critical to the future of all of condensed matter and materials physics. Nanoscience offers exciting new routes to create advanced materials and hierarchically assemble systems in a non-Edisonian manner. Utilizing the three basic tools of nanoscience—geometric confinement, physical proximity and self-organization—one can create materials by design. Confinement creates new properties from known materials via manipulation of dimensionality.

Proximity effects allow multi-component composites to behave as new materials that embrace properties that are often mutually exclusive and thus not found in singlecomponent systems. Self-organization utilizes rules of Nature to segregate and/or assemble systems on ultrasmall length scales that transcend the limits of present-day lithographic definition. Self-assembly shows promise of becoming a powerful approach to miniaturize structures in an affordable manner with very high uniformity. A challenge is to utilize self-organization, as is accomplished in biological systems, to create not just component materials, but fully functional, integrated systems.

One can imagine pulling an entire computer processor out of a test tube in the future. Achieving such goals certainly encompasses non-equilibrium synthetic routes, and the creation of multifunctional materials. Competing interactions and the preponderance of low-lying energetic states and quantum fluctuations help create the complexity that gives rise to unanticipated phenomena in magnetic nanosystems. We note that, during the exploration of magnetic surfaces, interfaces and multilayers undertaken a few decades ago, the discoveries of spin-valve read heads or magnetic random access memory was completely unanticipated. This realization underscores the fact that fundamental studies involving the creation of emergent matter is an important stimulant of advanced technologies.

The second grand challenge in nanomagnetism, beyond the creation of emergent matter, is the exploration of its magnetic properties. This task requires all levels of characterization tools. Essential in this is the utilization of the advanced user facilities that are available to the research community. This underscores an important synergy between materials creation and exploration that requires national investments both in synthesis and in major user facilities. Additionally, understanding emergent behaviour requires the talents of the theory community and the utilization of high-performance computing to simulate complex behaviour. This challenge has many facets. It includes: (i) defining the rules that govern self-organization; (ii) understanding spin dynamics and equilibrium states at the spatial and temporal limits of interest; (iii) understanding spin transport phenomena, spin-torque effects, spin accumulation, spin diffusion, etc.; and (iv) relating such effects to the detailed structure of the material. This challenge also involves understanding how to create entangled spin systems, and how to control the process as well as read the states.

A longer-term aspect of the challenge is to foster communication of information via spin currents and excitations without the flow of charge. This would be a novel approach to complement existing electronic circuitry that depends on charge flow.

Why create, explore and understand nanomagnetic emergent matter? An important reason is that the more detailed understanding which is the goal of basic science will enable the community to address the strategic needs of our society at large. It is vital to stimulate the economy and create new jobs by ushering in new technologies and innovations. New diagnostic tools and treatments for diseases will preserve health and wellbeing, as well as heal the ill and help the impaired.

New sensors will help ensure homeland security, and advanced materials and approaches will provide for national defence needs. It is important to create materials and harness phenomena that will advance transportation requirements, and it is imperative to take leadership in meeting national goals of energy independence. Additionally, all of these initiatives must be advanced in a manner that is respectful of the environment.

These are goals for the nanomagnetism community, for the nanoscience and nanotechnology communities, and for the larger materials research communities. The nanomagnetism community is in a particularly advantageous position to rise to these challenges because of the enormous potential derived from recent breakthroughs. Although magnetism is one of the oldest sciences known, it is a driving force in the new scientific era of nanotechnology. This statement is highlighted quite succinctly by examining the evolution of the giant magnetoresistance (GMR) effect, which made the transition from a laboratory curiosity to a technology driver within the short span of about ten years. From GMR, a host of new phenomena, materials, and potential technologies are blossoming. An overview of exciting aspects of magnetism research can be found in the collection of review articles entitled, Magnetism Beyond 2000.

MAGNETISM IN CONFINED GEOMETRIES

Magnetism in confined geometries is an active and vital area of research that will surely produce much new science and many applications in the next twenty years. In general, it is understood that systems with confined (or finite) geometries are those with dimensions on the order of the domain wall width, or of the mean free path of conduction electrons, i.e., nanometers.

Confined systems that exhibit novel properties often consist of dissimilar materials that include at least one magnetic component (ferromagnetic, antiferromagnetic, etc.). Geometries of interest at the nanoscale include thin film heterostructures, superlattices, and nanostructured, nanopatterned or granular thin films. A topical review of the fabrication and properties of ordered magnetic nanostructures was presented by Martin et al.

Arrays of nanoparticles and nanorods are discussed in Section 4. Although the future is difficult to predict, based on past experience, the area of confined magnetism is anticipated to produce breakthroughs in information storage, sensing and other spintronic applications. The main areas of interest in the study of confined magnetism include: (i) confinement and quantization by the finite size, and (ii) the proximity effect due to the leakage of the electronic wavefunction and attendant magnetization outside of the physical bounds of the nanostructure. Since magnetism depends on a variety of length scales, quantitative structural characterizations must be made at a similar variety of scales.

Examples of confinementinduced effects include vortex-state formation in magnetic nanodots, and modifications of the magnetic structure by surface defects. Proximity effect examples include coupling phenomena in Fe–Cr multilayers, induced magnetism in heterostructures built from nominally non-magnetic constituents, training effects in exchange-biased systems, and changes in the interfacial magnetization.

To study such phenomena from both applied and basic science perspectives one needs access to probes that enable the:

- Mapping of the magnetization at nanometer length scales and sub-nanosecond time scales.
- Study of buried layers and interfaces.
- Element-sensitive structural, chemical and magnetic characterizations.
- Pump–probe experiments carried out under photonic, electronic and magnetic stimuli.
- Monitoring and control of collective excitations such as phonons, magnons, polarons and excitons.

Ideally the above studies would have the potential to be conducted under specialized environments, including ultrahigh and ultra-low temperature, high pressure, and especially in variable magnetic fields up to 20 T. Furthermore, it is necessary to apply theoretical and numerical predictions of novel physics and effects of unusual geometries that extend beyond existing understanding. In particular, theoretical understanding of the interaction of X-rays with magnetic structures, especially in the presence of surfaces and interfaces and in conjunction with proximity configurations, is needed.

It should be stressed that the field of nanomagnetism is now beyond proof of concept. The time is thus ripe for applying sophisticated techniques to the solution of real physics-materials science problems that transcend simplifying assumptions. This requirement implies the ability to conduct detailed, quantitative studies that are correlated with materials growth processes and other physical property measurements. Selected examples will be used to highlight some of the scientific and technological issues that need to be explored in the area of magnetism in confined structures.

LATERAL SPIN STRUCTURES AND TRANSPORT

Understanding the transport of spins across interfaces in ferromagnetic (F) and non-ferromagnetic metallic (N) heterostructures is a fundamental challenge in the emerging field of spintronics.

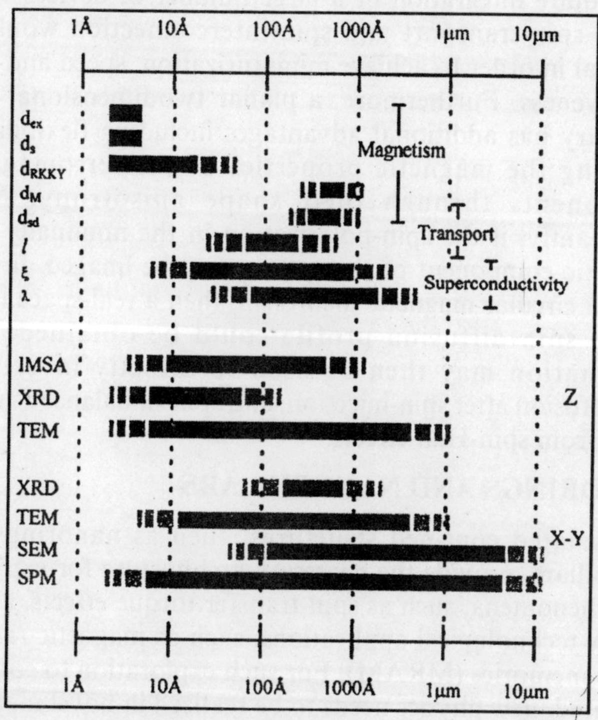

Fig. Important Length Scales in Magnetism and Superconductivity and Relevant Characterization Tools. The Density of Vertical Lines Indicates the Reliability and Range of Applicability of the tool at a Particular Length Scale.

The goal of spintronics, or magnetic electronics, is to explicitly utilize the spin of the electron instead of, or in addition to, the charge of the electron to communicate information. Spin-transport across interfaces gives rise to both the well-known GMR effect in F/N/F trilayers and F/N multilayers, and the recently discovered spin-transfer torque effects in nanoscale F/N/F trilayers. To date most work has focused on vertical, or stacked, heterostructures; however, it is also promising to explore lateral, or patterned, heterostructures

Since they offer additional degrees of freedom to tailor desired properties. Additionally, future spintronic applications will require integration of a large number of devices where lateral spin-transport and spin-interconnection would be essential in order to achieve miniaturization, speed and cost-effectiveness. Furthermore, a planar twodimensional (2D) geometry has additional advantages including flexibility in tailoring the magnetic properties of the ferromagnetic components through their shape anisotropy. More importantly, if the spin-polarization in the nominally non-magnetic component of the system may be imaged directly, i.e., via circular magnetic dichroism, then a real-space image of the spin-diffusion profile could be obtained. This information may then be used to directly investigate spindiffusion after spin-injection, and spin-imbalance as might occur from spin-Hall effects.

NANORINGS AND NANOPILLARS

Isolated confined structures, such as nanorings and nanopillars, provide the necessary architecture for exploring new phenomena, such as spin-transfer torque effects, as well as new technological applications, such as magnetic random access memories (MRAM). For such exploration to continue unabated, new physics needs to be realized in novel materials and structures made by new fabrication methods, and verified by advanced characterization tools.

A variety of small magnetic structures, including squares, rectangles, circular discs and rings have recently been studied. Such structures can form a vortex state, in which the magnetic flux curls to form an internally confined state.

However, there is a singularity at the centre of the vortex where the magnetization points out of the plane. In the vortex state of a circular disc, the magnetic moments along circles of decreasing radii must rotate ever more sharply. The singularity at the centre destabilizes the vortex state of the smallest circular discs that instead form into single-domain states. The fundamental obstacles of vortex singularities in circular discs are removed in magnetic nanorings.

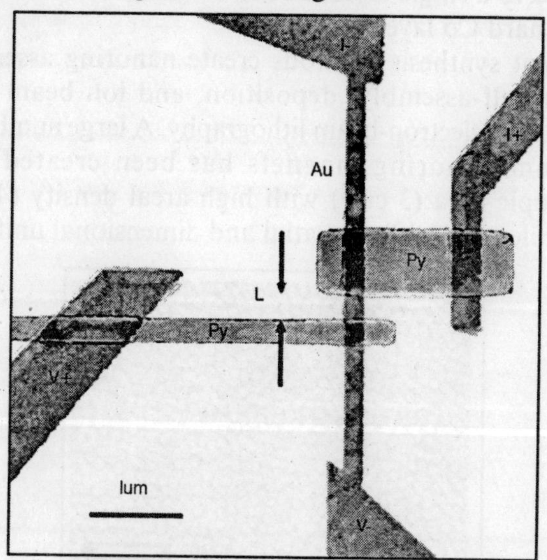

Fig. Scanning Electron Microscopy of a Permalloy/Au/Permalloy Lateral Spin Valve

Not only can a true vortex closure state be realized in this geometry, but magnetic nanorings have other unique attributes. A magnetic nanoring can acquire two distinct vortex states of opposite chirality, with magnetization circulating either clockwise or counterclockwise, in addition to bi-domain and other states containing domain walls. Each nanoring can therefore potentially store information as one or more magnetic bits. Magnetic rings have been created recently using electron-beam lithography. The smallest rings reported are 90nm in diameter and 30 nm in width.

The electron beam lithography technique is typically used to pattern relatively small areas, and the small number of rings in each array defies most magnetic characterizations except for the magneto-optic Kerr effect (MOKE) technique, resistance measurements, and in some cases, magnetic force microscopy. As an example, giant magnetoresistance measurements of Co/Cu/NiFe rings show that the soft NiFe layer reverses by a different mechanism in the multilayer compared to a single layer, due to magnetostatic interactions with the hard Co layer.

Recent synthesis methods create nanoring assemblages based on self-assembly, deposition, and ion beam etching without using electron-beam lithography. A large number (109) of 100-nm nanoring magnets has been created over a macroscopic area (3 cm^2) with high areal density (45 rings/mm^2) while maintaining spatial and dimensional uniformity.

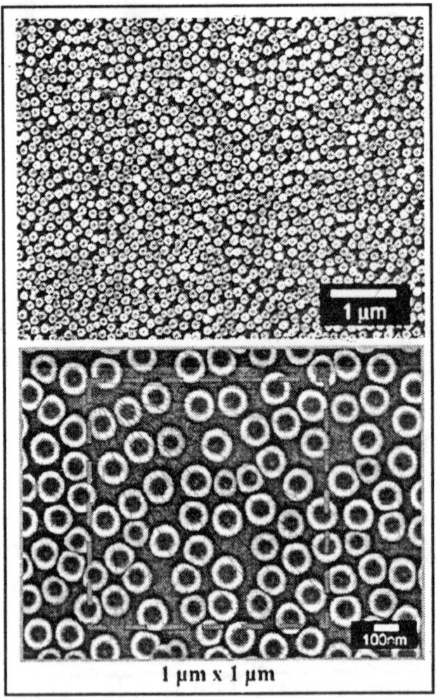

Fig. SEM images of Co nanorings with an inner diameter of 100nm (top) with a high areal density of 45 rings/mm^2 as shown in the lower.

Because of the large number of near-uniform rings, the switching characteristics can be readily measured, by means of a vibrating sample magnetometer (VSM), and compared with micromagnetics simulations. Since the lateral dimensions of the nanorings are in the 100-nm range, traditional magnetic force microscopy (MFM), with a lateral resolution of 30–50 nm, is no longer adequate. Spin-sensitive microscopy is needed using X-ray photons or other means with a lateral resolution < 3 nm. Another example of a confined structure with novel magnetism is nanopillars. Current-induced magnetic reversal has been demonstrated in Co/Cu/Co multilayer spinvalve nanopillars with particular relevance to MRAM devices. These pillars, synthesized by a novel fabrication process, possess lateral dimensions on the 100–200 nm scale.

Depending on the applied field and its frequency, the current can cause in these pillars either magnetic reversal or precession, accompanied by the emission of microwaves at distinct frequencies. Key to the full understanding of this effect is the determination of the dynamic behaviour of the magnetization of the various layers in these structures on very short timescales and with high spatial resolution.

Confined Magnetic Nanosystems for Magnetic Information Applications

Magnetic storage has played a key role in audio, video and computer development since its invention more than 100 years ago by Valdemar Poulsen. In 1956 IBM built the first magnetic hard disk drive featuring a total storage capacity of 5Mbytes at a recording density of 2 kbits/in2. Since then the density of bits stored on a surface of a disk has increased by _50 million fold to densities in 2005 of _100 Gbits/in2 and has been doubling every year over the past five years. At such densities, the bits must be positioned on the disk with nanometer resolution.

Presently, magnetic disk drives are based on longitudinal recording systems where the magnetization of the recorded bit lies in the plane of the disk. These systems contain a recording head composed of a separate read and write element, which flies in close proximity to a granular recording medium, as illustrated in figure. The recording media signal-to-noise (SNR) ratio needed for high-density recording is achieved by statistically averaging over a large number of weakly interacting magnetic grains per bit.

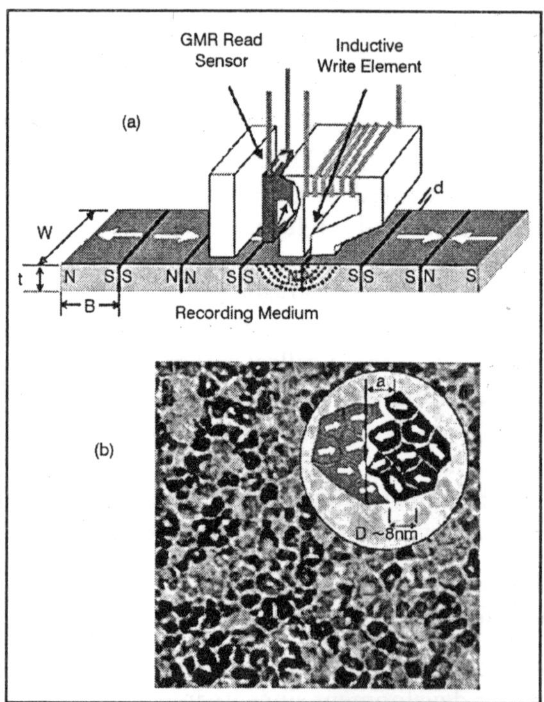

t is the medium thickness, W is the width of the recorded track, B is the bit size, and d the height the head flies above the medium. (b) Plan-view electron microscope image of a CoPtCrB magnetic alloy longitudinal recording layer. The amorphous grain boundaries are seen as white and the average grain size is 8.5 nm.

Shown in the inset is a magnification of a transition region showing the interplay of the medium magnetic properties with the microstructure where D is the media grain size and the transition parameter a approximates the width of the magnetization reversal region. (c) Schematic illustration of a perpendicular recording system with a read and write head flying above the recording medium. The magnetization of the media is perpendicular to the films and a soft magnetic underlayer is below the medium to enhance the perpendicular write fields from the head. (d) Plan-view electron microscopy image of a perpendicular CoPtCr-oxide recording medium used to record data at 230 Gb/in2. The average grain size is 6.3nm with a coercive field of 6.4 kOe and a magnetic energy per grain of 75 kB T.

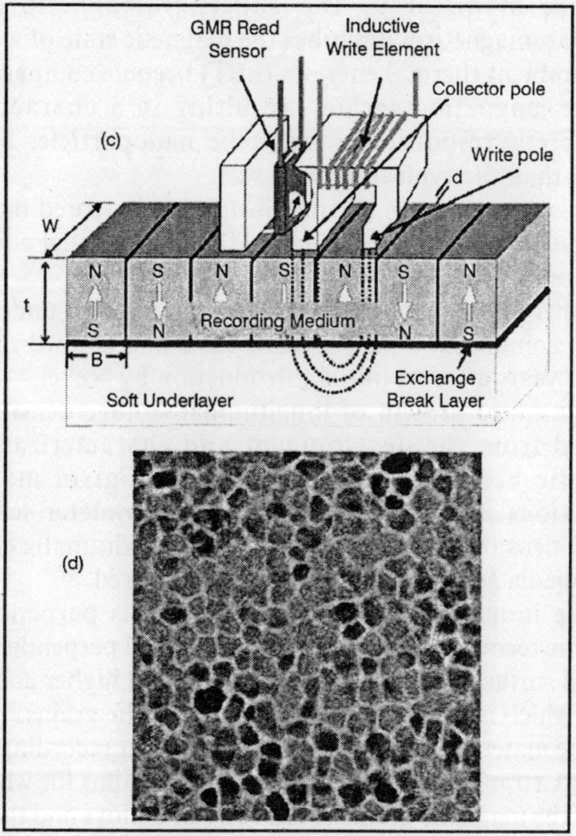

Fig. Schematic Illustration of a Longitudinal Recording System with a Read and Write Head Flying Above the Recording Medium.

The granular microstructure limits the magnetic correlations to length scales comparable to the grain size and allows information to be written on a finer scale than possible in a homogeneous magnetic film. This structure is currently achieved in CoCrPtB alloy thin films. When grown at elevated temperatures, these alloys phase segregate into high moment magnetic grains surrounded by non-magnetic boundary regions as shown in the TEM image in figure 4b. The continued evolution of recording technologies towards higher areal densities is limited, in part, by enhanced thermal fluctuations that are manifest in the superparamagnetic effect. Superparamagnetism describes the magnetic state of a system when ambient thermal energies (kBT) become comparable to volume magnetic energies, resulting in a characteristic anhysteretic response for magnetic nanoparticles of sizes smaller than the typical domain size.

This effect creates thermal instability in stored magnetic information and produces thermal fluctuations that add noise to sensors.

Superparamagnetism is an issue that is well-understood yet still commands a high level of basic and applied research over 50 years after its initial introduction by Ne′ el.

Continued growth of longitudinal storage densities has resulted from the development and characterization of magnetic heterostructures that help control magnetic interactions and correlations on the nanometer scale. As storage densities continue to increase more dramatic changes in the media structure and system are required.

The industry is now moving towards perpendicular magnetic recording where the bits are stored perpendicular to the film surface. This allows for thicker and higher coercivity media which increase the energy stored in the grains.

The materials of choice for perpendicular recording media of CoPtCrtransition metal oxide composite films for which the core of the magnetic grains consist of CoPt alloys and the oxide segregates to the grain boundary to isolate the grains. Beyond perpendicular recording, there are also new developments in thermally assisted recording, patterned media, and self organized magnetic nanoparticles.

Determining the potential of these new structures (and the opportunities for magnetic X-ray studies) will occupy much research talent and require the capability in the future of characterizing magnetic materials at the nanometer spatial scale and sub-nanosecond temporal scale needed for future storage devices.

Patterned Media from Block Co-Polymers

The top-down synthesis strategy of lithography employing self-assembled block copolymers allows controlled fabrication of confined nanostructures with exceedingly regular features.

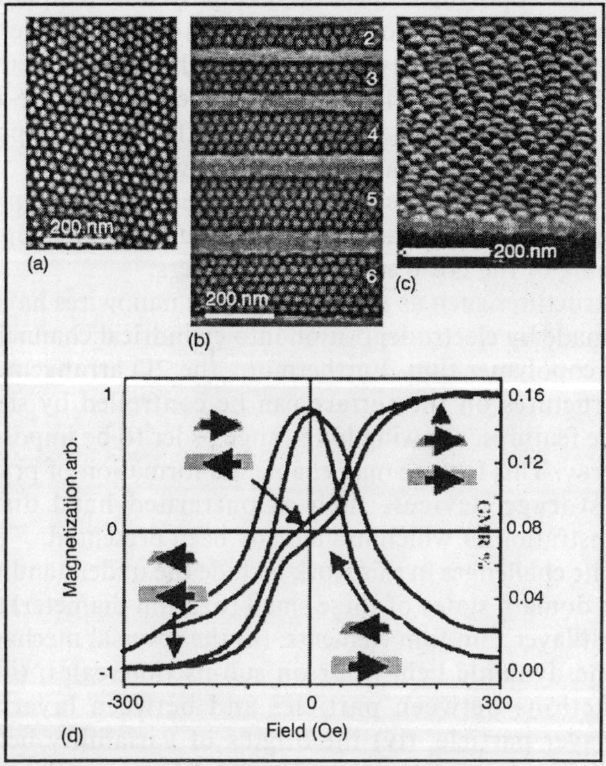

Fig. (a) Self-assembled block copolymer on a smooth substrate, showing lack of long range order. (b) In shallow grooves, the polymer forms long range ordered structures. (c) Array of Co 'dots' made by block copolymer lithography. (d) Magnetization (black) and magnetoresistance (blue) of an array of 35-nm diameter dots made from a CoFe (3.3 nm)/Cu (6 nm)/NiFe (4.5 nm) multilayer.

Such structures are of great interest for patterned media applications with high densities (e.g. periodicity o50nm over large substrate areas). Diblock copolymers consist of polymer chains made from two chemically distinct, immiscible polymeric materials that are joined at one point. Self-assembly of these polymers allows formation of small-scale ordered patterns whose size and geometry depend on the molecular weights of the two types of polymer and their interaction.

Block copolymers have been employed as templates for the formation of magnetic nanoparticles by selective removal of one of the polymer types and use of the resulting template to pattern a magnetic film. The magnetic nanoparticle arrays made by this technique show strong effects of interparticle interactions as well as a magnetoresistance.

The multilayer structure is preserved during the fabrication process, and the data illustrate the separate switching of the NiFe and the CoFe layers.

Structures such as aligned magnetic nanowires have also been made by electrodeposition into cylindrical channels in a block copolymer film. Furthermore, the 2D arrangement of the structures on the surface can be controlled by shallow surface features, allowing long-range order to be imposed on the array. This feature may enable the formation of practical data storage devices, such as patterned hard disks, a demonstration of which has already been presented.

The challenges in this work include the understanding of: (i) the domain states of these small (o25-nm diameter) single or multilayer thin film elements; (ii) the reversal mechanisms and the dynamic behaviour on sub-ns timescales; (iii) the interactions between particles and between layers in a multilayer particle; (iv) the origins of variability between nominally identical particles; and (v) how to relate these phenomena to the array geometry, the microstructure, the edge, interface and surface structure, the presence of surface oxides, and the shape and magneto-crystalline anisotropy.

MAGNETIC RANDOM ACCESS MEMORIES (MRAM)

MRAM is a new technology that employs a magnetoresistive device integrated with standard silicon-based microelectronics, resulting in a combination of speed, nonvolatility, and endurance not found in other memory technologies. The first generation of commercial MRAM went into production in 2006. Due to its unique combination of attributes, MRAM has the potential to be a universal memory that replaces several memory types.

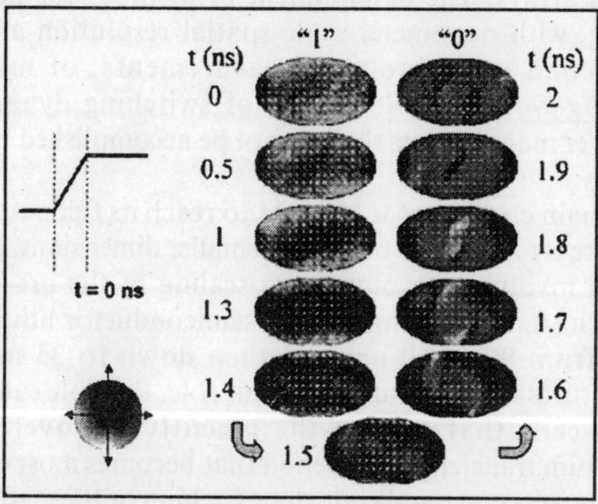

Fig. Time-dependent micromagnetic simulation with random thermal fluctuations of a 4nm NiFeCo MRAM bit patterned as a 0.6 X 1.2 mm^2 ellipse. The current pulse was given a 2 ns rise time, and the element reversed within 2 ns.

Since it is based on sub-micron magnetic devices, continued development of MRAM to smaller dimensions depends on advanced research in nanometer-scale magnetic structures. Presently, materials research areas that are important to future generations of MRAM include: high-polarization materials, alternate tunnel-barrier dielectrics, and fundamental studies of specific electrode/dielectric combinations. Progress in these areas could enable higher performance MRAM at smaller dimensions by providing higher signal levels. For example, improved materials are needed to reduce variations in the tunnel barrier that in turn cause variations in the bit-to-bit resistance.

Writing data to MRAM cells involves switching the magnetic moment of the storage layer between two stable states. A micromagnetic simulation of an MRAM bit switching. Comparable experimental images on real MRAM bits would improve our fundamental understanding of magnetization reversal in these complex structures and help minimize the bit-to-bit switching variations observed in Mbit MRAM arrays. The development of methods for magnetic imaging, with nanometer-scale spatial resolution and sub-nanosecond time-resolved measurements, of magnetic switching would enable studies of switching dynamics in multilayer magnetic bits that cannot be accomplished by other methods.

A main challenge for MRAM to reach its full potential in the future is to continue to scale to smaller dimensions. Initially this will involve an evolutionary scaling of the present-day approach. As the leading edge of semiconductor lithography moves from 90 to 60 nm and then down to 35 nm, new architectures may play an important role. Possible candidates are bit-cells that exploit the recently discovered spin momentum transfer phenomenon that becomes most effective at dimensions o50 nm. While the switching of these structures has already been demonstrated, much work remains in materials research and fundamental understanding of nanomagnetics. Other new materials and magnetic device structures that enable switching at low currents with good data retention and higher read signals would have a large impact on the evolution of MRAM and its incorporation into the mainstream memory market.

Interaction between Spin-Current and Nanomagnets: Spin-Torque

In a GMR or tunnel MR (TMR) device, spin-torque refers to the inverse action of a spin-polarized transport current on the nanomagnet traversed by such a current. This interaction becomes important for devices o100 nm or so in size. It manifests itself as a currentinduced magnetic excitation and/ or magnetic reversal of the nanomagnet. The interaction is important to understand, especially for applications, including memory technology, microwave generation, and recording head performance. For fundamental science, this is a new interaction between a spin current and a ferromagnet. It introduces a new mechanism for dynamic excitations. It also connects the magneto-transport problem to that of a magneto-dynamics problem, affecting our understanding of magnetic damping, spin-transport and spin-pumping in extended spin-circuits.

An illustration of the spin angular momentum transfer process. When a charge current passes through a spin-valve (or magnetic tunnel junction) stack as shown on the lower left of panel (a), the current becomes spin-polarized.

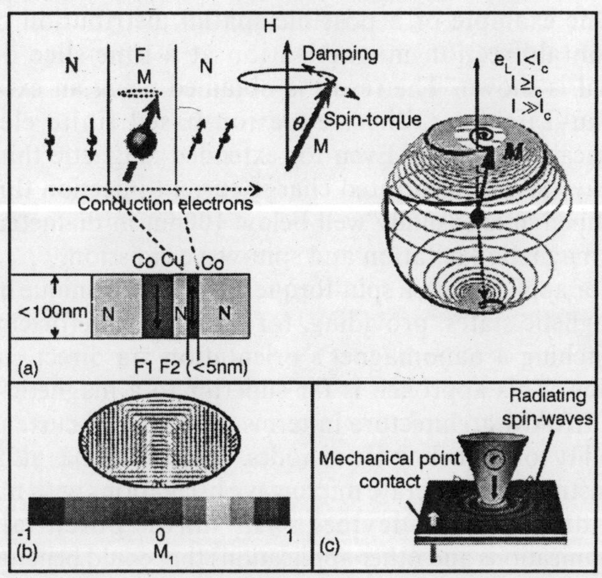

Fig (a) Illustration of the spin angular momentum transfer process. (b) A nanostructured thin film pillar structure with a possible spatial distribution of the horizontal-direction magnetization at a time-sliced during reversal is shown using an extended Landau–Lifshitz– Gilbert equation-based finite element numerical simulation (c) A concentrated charge current injection through a confined area can cause magnetic excitation and spin-wave emission,

The magnetic free layer (F2) will absorb some transverse spin angular momentum when repolarizing the spin-current, leading to an effective torque (spintorque.) The direction of this torque can be parallel or antiparallel to the damping torque, depending on the sign of current and spin-polarization. When the direction of the spin-torque opposes the damping, a net increase in precession cone-angle may result if the amount of current supplied exceeds a certain threshold, leading to a precession state that can result in magnetic reversal as illustrated in the right side of panel. A nanostructured thin film pillar structure, for which the magnetic state during spin-torque excitation and reversal can have a complex spatial distribution.

One example of a possible spatial distribution of the horizontaldirection magnetization at a time-slice during reversal is shown. The result is obtained using an extended Landau–Lifshitz–Gilbert equation-based finite element numerical simulation. Even for extended magnetic thin film multilayers, a concentrated charge current injection through a confined area (usually well below 100nm in diameter) can cause magnetic excitation and spin-wave emission.

For applications, a spin-torque introduces a unique handle on magnetic states, providing, for example, an efficient way of switching a nanomagnet's orientation via direct current injection. This approach is far superior to a magnetic-field-write MRAM architecture in terms of low write-current and scalability to well below 45nm nodes. Spin-torque has also been demonstrated to generate microwave oscillations with narrow linewidths. Such devices will have potential for communications and other applications that could benefit from a compact and tunable microwave source. The spin-torque interaction is also affecting the magnetic recording head's dynamic performance in a profound way. Head devices, both spin-valves and tunnel junctions, will need to be optimized taking into account the effect of measurement current on the magnetodynamics and magnetic noise characteristics. Similarly, a spin-polarized current traversing a magnetic domain wall can also interact with it, causing domain wall motion, providing another interesting way of manipulating magnetic states of nanomagnetic structures.

Such magnetic states inducible by a spin-polarized current

tend to have unique dynamics different from field-induced dynamics. High intensity, pulsed and tunable X-rays from modern (third generation) synchrotron sources do have the spatial and temporal resolution for the direct observation of these novel magnetic structures and for quantitative understanding of the element-specific magnetic dynamics.

SPIN DYNAMICS

A significant class of experiments utilizes optical pump–probe techniques to address spin dynamics in confined, interacting systems such as nanoparticle arrays. In these techniques, a sample is pumped by a pulse that generates an excitation. After a set period of time, a probe pulse characterizes the effects of the pump by means of transmission or reflection studies. In this manner spatial resolution of o300 nm can be achieved with ultraviolet light. In cases where fast magnetic field pulses are generated with a strip-line, the practical time resolution is of order 50 ps. This technique permits the study of magnetic excitations at long wavevectors and frequencies o10 GHz. Phenomena such as ultrafast switching, spin-wave localization, and vortex gyrotropic modes in individual nanoparticles have been predicted and explored experimentally. Most of these systems are characterized by strong inhomogeneities in the internal magnetic field and/or in the magnetic microstructure.

An example is shown for a magnetic vortex. This type of structure forms in cylindrical nanoparticles, in which the magnetization curls around the perimeter of the disk to minimize the magnetostatic energy. A distinguishing feature of these structures is the existence of the vortex core in which the magnetization is forced to point out of the plane. Although this core cannot be resolved optically, it has a profound influence on the magnetization dynamics. There is a node in the response at the centre of the disk. Unfortunately, the spin dynamics in this core region cannot be readily explored using any existing experimental techniques. A similar technological limit has been encountered in recent experiments on ferromagnetic nanowires.

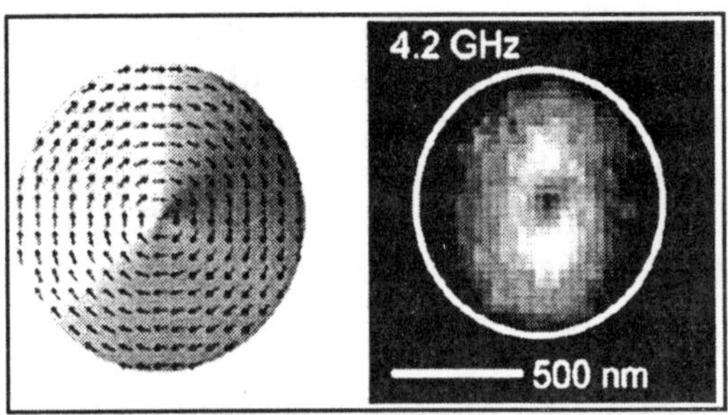

Fig. A Magnetic Vortex Schematic (left) and an Experimental Image Based on the Work by Park and Crowell of a Spin-wave Mode in a 1-mm Diameter Disk at 4.2 GHz.

Note the dark spot (node) in the centre of the experimental image. Accessing this region requires a probe with spatial resolution on the order of the exchange length (5 nm.).

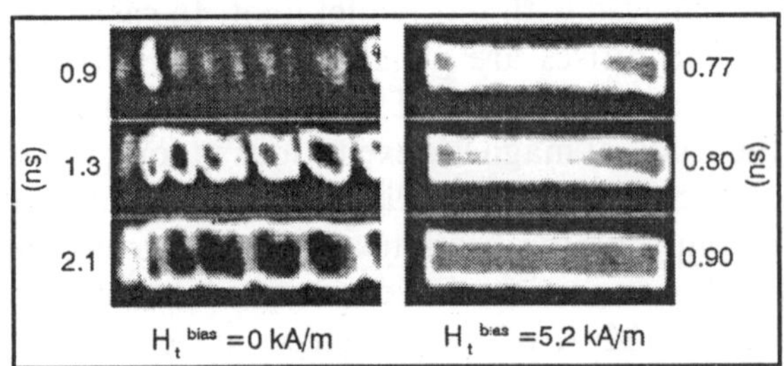

Fig. Spatial Profiles of the Easy axis Magnetization Component Measured at Different Time Delays after a Magnetic Switching Pulse was Applied.

The magnetization reversal mode is by incoherent rotation on the left, where the switching field is applied directly opposite to the initial direction of the magnetization. In the column on the right, the reversal is by quasi-coherent rotation, on account of the initial direction of the magnetization having been tipped away from the geometric and magnetocrystalline easy axis by a small transverse bias field. The field of view of each frame is 12 X 4 mm, and contains the entire 10 X 2 mm sample. The dynamic magnetic structure is full of details (spin waves, domain walls, vortex cores) on the nanometer scale and invisible to the magnetooptical probe used for this experiment.

As the confining dimension decreases to the order of the exchange length (_ 5 nm), the fundamental character of the excitations should change. A time-resolved probe that enhances spatial resolution from the present level of _100 too50nm would improve our measurement capabilities, while a probe with 5-nm resolution would enable the exploration of fundamentally new physics. In order to fully exploit the advantages of improved spatial resolution, there should be an accompanying increase in the temporal bandwidth. If, for example, a measurement could be performed with 1THz of bandwidth (i.e. 1 ps pulses) and 5-nm spatial resolution, then it would be possible to probe excitations that presently fall in the unexplored window of momentum space between time-resolved optical experiments and inelastic neutron scattering. This advance would permit time-resolved studies of quantum spin dynamics.

It should be stressed that these measurements rely on phase-coherent repetitive measurements. The data are obtained stroboscopically (utilizing "boxcar" averaging) and hence can only capture deterministic phenomena. This limitation means that currently laboratory experiments are not sensitive to many possible stochastic effects. Thus single-shot, time-dependent magnetic measurements, carried out at sub-nanosecond time scales and at nanometer length scales, would be important to characterize a greater variety of phenomena and should thus emerge as one of the important future nanomagnetism research directions.

The ultimate goal for magnetic dynamics studies is to achieve 1 nm spatial resolution and o10 ps temporal resolution. Adoption of a sub-nanosecond (_100 ps) time resolution goal would be insufficient to probe precessional frequencies of 10's GHz, nor would it be able to track magnetization reversal by incoherent rotation which, at _1 km/s (μ 1 nm/ps), is not the fastest of magnetization reversal mechanisms. Hence, there is a need for probes that can yield information about single-domain particles that ultimately cannot be described in the single-spin approximation. This is the frontier at which the traditional Landau–Lifshitz–Gilbert model for dynamics becomes wholly inadequate, and at which the crossover to even smaller (cluster) magnetic systems naturally occurs. Synchrotron X-ray methods may hold special promise for such investigations because of their ability to yield elementspecific information and because of the potential for combining high spatial and temporal resolution.

Theory and Computation of Confined Magnetic Systems

In the computational arena, magnetism and magnetic materials have been traditionally studied with phenomenological models. These models either work well or must be supplemented by new terms in the model to account for unexplained effects. While this empirical approach offers useful insight into the underlying science and provides benchmarks for the analysis of experimental data, it has its limitations. Chiefly the model may not be able to explain characteristics that depend on the details of the system at the nanoscale and may not have predictive capabilities. Magnetic properties at interfaces and surfaces, which make up a large fraction of nanostructured and confined materials, are quite different from the bulk systems upon which many simple models are built. Model Hamiltonian approaches to magnetic nanostructures thus have limitations: either the Hamiltonian is too simple, based on bulk parameters that miss the essence of nanomagnetism, or the Hamiltonian is too complex and consists of many terms with unknown parameters that is unsolvable.

It is fortunate that, the theory of magnetism and magnetic materials is already on a rather firm footing. Ab initio electronic structure methods are capable of predicting basic materials properties, such as the magnetization, magnetic anisotropy, the exchange energy, and the Curie or Ne´ el temperatures. As computer performance increases and computational methods become more efficient, these first-principle methods can approach the point where substantial portions of nanostructures can be simulated without the need of fitting parameters.

These methods will provide information beyond what is

presently accessible experimentally on the following aspects: (1) atomic-scale magnetic structure in the interfacial and surface region of nanostructures; (2) nanomagnetism at short time and small length scales, and (3) understanding of the interaction of X-rays with magnetic nanostructures. It is important that theoretical calculations yield specific predictions that are experimentally accessible. Advanced models that are built on ab initio theory and experimental observation will lead to systematic advances in our understanding of nanomagnetism. The weaving together of theories that cover discrete spatial and temporal realms into a single hierarchical computational package remains a major challenge. The exploitation of quantum computation also appears on the horizon.

CLUSTER MAGNETISM

Cluster magnetism describes magnetic phenomena found in nanoscale systems that are heterogeneous in all three spatial dimensions. The heterogeneity may have physical or chemical underpinnings and can exhibit a number of degrees of complexity. The magnetic character of cluster systems depends on both the intrinsic properties of the individual clusters as well as on their interactions. These interactions determine the type of ordering within the system: the system heterogeneity may be regular, exhibiting either short- or long-range order, such as that found in selfassembled arrays of nanoparticles and molecules. Conversely, the system heterogeneity may be random, as that found in spin glasses, spin ice and granular systems. An additional complexity is derived from the nature of the space between the clusters. The inter-cluster space may consist of organic polymer ligands, a matrix phase of insulating or conducting character, or vacuum. All of these factors ultimately determine the magnetic response and transport properties of the system.

The study and application of cluster magnetism is of great value from a number of perspectives. From a basic physics point of view, nanoparticle arrays and molecular magnets are test-beds for understanding the effects of nanostructuring, magnetic coupling, and of correlations from both static and dynamic perspectives. Superparamagnetism, discussed in Section 3.3, is an important phenomenon in nanoparticle arrays. As the scale of the system becomes smaller, superparamagnetism emerges at higher temperatures. This phenomenon has important ramifications for technological devices employing magnetic cluster systems, such as magnetic resonance imaging (MRI) contrast agents and recording media.

With rational synthesis it is possible to create uniform model nanoparticle systems that will foster understanding of emergent properties, especially in concert with theory and modeling. The science and practice of nanoparticle synthesis is steadily moving forward, to create a variety of shapes and geometries with unusual properties. While perhaps the best-known application is magnetic recording media, cluster magnetism may also expand the development of ultra-strong magnets for highly efficient motors, enable multifunctional sensors, and tailor superior soft magnets for electrical distribution. From a biological or biomedical point of view, magnetic clusters often cover the same size range as do biomolecules.

This fortuitous coincidence cultivates interdisciplinary science between physics and biology as well as fosters new applications in pharmacology, medical treatment protocols and imaging. Cluster magnetism review articles include the theory of fine magnetic particles, magnetic and transport properties of fine magnetic particles, and molecular magnets. The objective of this section is to highlight forefront scientific issues regarding cluster magnetism and associated technological issues. In the following sections, recent advances are described in a select group of subtopics.

MOLECULAR MAGNETISM

Within the topic of cluster magnetism are organic-based or molecular nanomagnets, also referred to as single molecule magnets (SMM), which are a class of materials that exhibit a broad range of both conventional and new physical phenomena. SMM are distinct from inorganicbased nanoparticles by virtue of their constituent organic ligands that allow complete localization of the magnetic moments in the ferromagnetic atomic component. Molecular magnets contain a very large number (_Avogadro's) of magnetic molecules that are nominally identical, providing ideal laboratories for the study of nanoscale magnetic phenomena. Interest in these materials has grown dramatically in the last several years, owing to their possible use for high-density information storage, as well as the possibility that some members of this materials family could provide the qubits needed for quantum computation.

An advantage of SMM over other potential quantum computer materials is that the nanoscale size of the individual molecules provides the ability to pack _1020 clusters into a cubic centimeter. Although there are no commercial devices based on organic-based magnets at present, it is anticipated that applications will be developed. For example, inexpensive, disposable, large-area MRAM made from self-assembled organic-based magnets and semiconductors could complement conventional inorganic- based electronic and magnetic materials.

SMM typically consist of monodispersed, nanoscale clusters of 2–15 magnetic core ions embedded in non-magnetic ligand groups. As one prototypical example, the full chemical formula for Mn12-acetate is ($[Mn_{12}O_{12}(CH_3COO)16(H_2O)_4]$ – $2CH_3COOH$ – $4H_2O$). These units are in turn packed into large single crystalline arrays.

The magnetic core ions interact mainly through Heisenberg exchange interactions with intramolecular magnetic interaction strengths in the range 1–100 K, whereas the much weaker intermolecular magnetic interactions are _10mK in strength. With molecular clusters of large total spin of 10 mB, Mn12 and similar molecular assemblages containing Fe8 exhibit properties that straddle the border between classical and quantum magnetism. These clusters are magnetically bistable at low temperatures and exhibit macroscopic quantum tunneling between up- and downspin orientations, as well as quantum interference between tunneling paths. Prototypical of the class, Mn12- acetate contains magnetic clusters, that are composed of 12 Mn atoms coupled by super-exchange through oxygen bridges to provide a S μ 10 spin magnetic moment that is stable at temperatures of the order of p10 K.

The nominally identical, weakly interacting clusters are regularly arranged on a tetragonal crystal lattice. As illustrated by the double well potential of figure. 11, strong uniaxial anisotropy yields doubly-degenerate ground states in zero field and creates a set of excited levels corresponding to different projections of the magnetization mS, with mS μ +10, +9y 0, of the total spin along the easy c-axis of the crystal. Magnetic relaxation proceeds in these systems by spin reversal via thermal excitation over the anisotropy barrier and/or by quantum tunneling across the potential barrier. Below the blocking temperature TB_3K, a series of steps appear in the magnetization M versus field H curves, indicating enhanced relaxation of the magnetization by tunneling whenever levels on opposite sides of the anisotropy barrier coincide in energy.

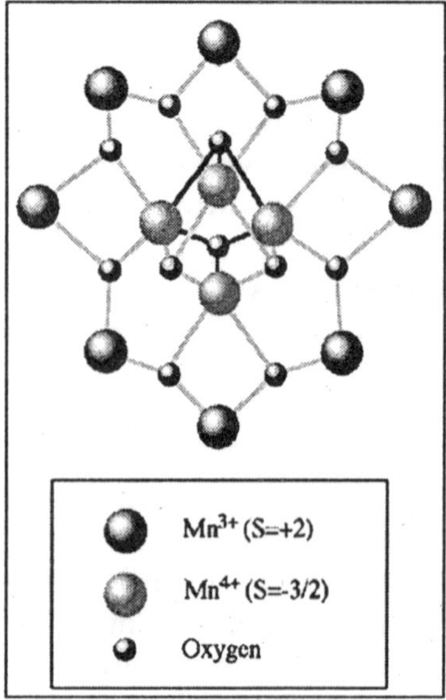

Fig. Schematic Depiction of the Magnetic Clusters in Mn12-acetate.

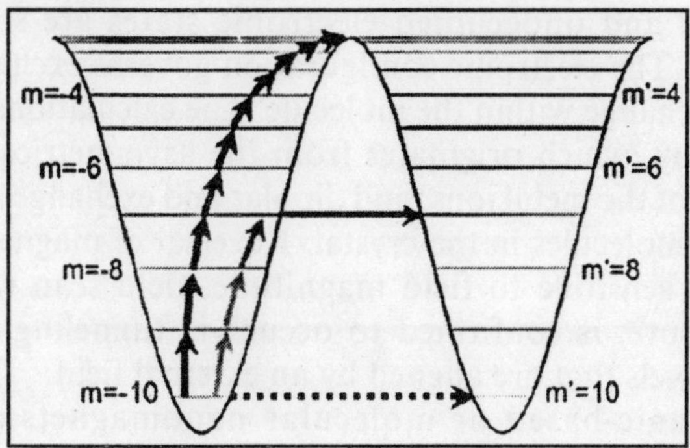

Fig. Double Well Potential of Mn12-acetate Weakly Interacting Clusters with Doubly-degenerate Ground States in Zero Field.

Magnetic relaxation proceeds in these systems by spin reversal via thermal excitation over the anisotropy barrier and/or by quantum tunneling across the potential barrier.

Definitive confirmation of the quantum mechanical nature of the magnetic relaxation is provided by the demonstration in Fe8 of quantum mechanical interference between different tunneling paths. Because of their atomic level monodispersity, the energy levels of the quantum mechanical spin states are extremely well-defined, and Zeeman splittings in an external field have been accurately mapped. By choosing a field where two levels cross, it becomes possible to prepare a quantum mechanical superposition state, i.e., an entangled state that is necessary for making qubits for a quantum computer. High spatial resolution polarized X-ray probes could be extremely useful for futures studies of such systems.

In another recent example, a number of SMM with magnetic cores consisting of just 2–4 magnetic ions have been prepared and characterized. In these molecules, the transition-metal 3d-atoms (TM) link to each other via oxygen giving rise to a TM–O–TM super-exchange interaction with magnetic characteristics that depend on the TM–O–TM angle. The ligands also provide an opportunity for manipulating the nature and/or the strength of the exchange interaction. Results from ab initio electronic structure calculations that identify the electronic configuration indicate that the density of the occupied and unoccupied electronic states are strikingly different. The electronic configuration governs exchange and super-exchange within the molecule. The calculations include anisotropy, which originates from the asymmetrical spatial location of the metal ions, and dipolar and exchange coupling between molecules in the crystal. Reversal of magnetization, which is sensitive to field magnitude, field scan rate, and temperature, is confirmed to occur via tunneling between energy levels that are aligned by an external field.

Organic-based or molecular nanomagnets are also important in the biological systems. For example, hemoglobin is an iron-containing respiratory pigment protein in blood that transmits oxygen from the lungs into tissues. This protein, which may be considered as a molecular nanomagnet, possesses a Fe–O bond length that is longer in the high-spin state than it is in the low-spin state. This feature means that the two spin states have different structures and therefore possess diverse chemical characteristics.

The benefit to society of successful development of organic molecular magnets is substantial. Implementation of new technologies will open up innovative means of addressing many needs including in electronics, energy consumption, health care delivery, and consumer packaging.

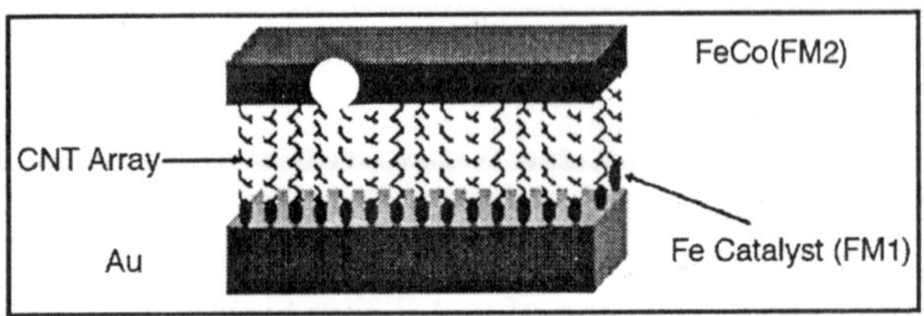

Fig. Schematic Illustration of a Spin Valve Device Employing Carbon Nanotubes.

Unlocking the potential of molecular magnets requires focused experimental and theoretical research. Topics for clarification include investigation of the existence of robust spin injectors, such as organic semiconductors and embedded inorganic magnetic nanoparticles, the possibility of magnetic organic, fully-spin-polarized semiconductors, and of photo-induced magnetism. For example, a spin-valve device in which the non-magnetic layer consists of an oriented carbon nanotube (CNT) array. These devices are intriguing, as many organic SMM arrays are transparent visible light, setting the stage for photo-induced magnetic phenomena.

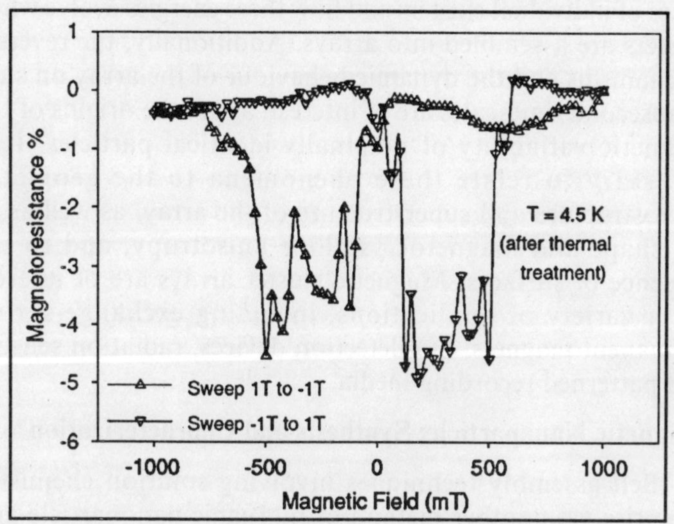

Fig. Magnetoresistance—defined as qR = R (H µ 1 T) Versus Magnetic Field for FeCo/Carbon Nanotube Array/(Fe Nanoparticle) Spin Valve Device at 4.5 K.

The negative magnetoresistance is obtained for a sample for which the FeCo layer was allowed to oxidize under ambient conditions for several weeks to form an oxide. The formation of an oxide at the interface between FeCo layer (as a result of aging and thermal treatment) and the CNTarray is the likely origin of the reversal of sign of the spin valve effect. Note the length of the CNTs between the Fe nanoparticle and the FeCo layer is _7 mm.

Open questions include the nature of the interfaces in SMM arrays, and clarification of the interand intra-cluster magnetic exchange coupling. Specifically, questions are centered on how electron spins in these systems traverse interfaces without losing their orientation.

INORGANIC MAGNETIC CLUSTERS

An area of great fundamental interest and critical technological need is the fabrication and magnetic property characterization of inorganic magnetic cluster arrays. Scientific challenges include understanding interactions and domain states of individual clusters and how these energies evolve when clusters are assembled into arrays. Additionally, the reversal mechanisms and the dynamic behaviour of the array on sub-nanosecond timescales are of interest, as are the origins of the magnetic variability of nominally identical particles. It is necessary to relate these phenomena to the geometry microstructure and superstructure of the array, as well as to the shape and magnetocrystalline anisotropy, and to the presence of surfaces. Magnetic cluster arrays are of interest for a variety of applications, including exchange-spring permanent magnets, bio-detection devices, radiation sensors and patterned recording media.

Magnetic Nanoparticle; Synthesis and Characterization

Self-assembly techniques involving solution chemistry comprise yet another method of inorganic nanoparticle and array synthesis. Highly uniform, monodispersed, surfactant-coated nanoparticles form stable dispersions in alkanes or toluene. The consequences of this monodispersity is that the particles can then self-assemble into arrays, just as atoms join to form close-packed fcc or hcp lattices when the solvent is evaporated, with slower growth processes leading to larger structural coherence lengths.

The size of the nanoparticles and the interparticle distance may be controlled via chemical variations. Super-para-magnetic-to-ferromagnetic and insulator-to-metal phase transitions are expected as the interparticle distance is varied. Important issues related to self-assembly include understanding what kind of magnetically interesting structures can be formed, and what are the forces responsible. Interactions between particles, and between particles and a surface within a fluid, are understood on the micron scale but not on the nanoscale, where, for example, the notion of an electrostatic double layer surrounding a particle breaks down. There is a need for both experimental work and theoretical modeling in order to refine the understanding of these interactions. An important technological goal is the self-assembly of defect-free nanoparticle arrays over macroscopic length scales. Such arrays could be useful for data storage media, and can form more complex structures that might be useful in biomedical applications.

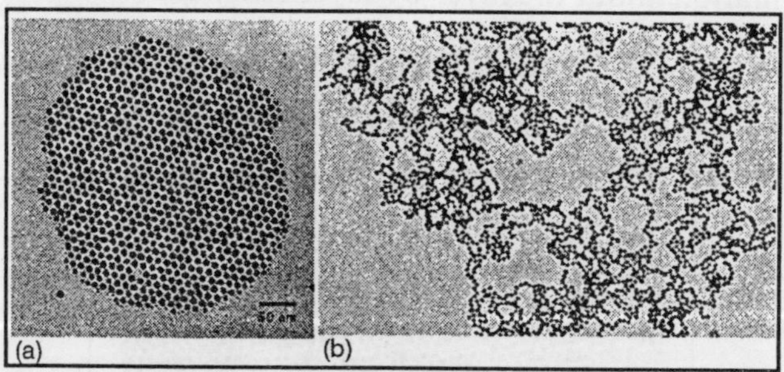

Fig. (a) TEM image of 9 nm Co nanoparticles in an array self-assembled from a hexane dispersion. (b) TEM image of _10-nm Fe3O4 nanoparticles in a magnetic gel self-assembled from an aqueous dispersion. The forces guiding self-assembly on this length scale are not yet sufficiently understood to make arrays of magnetic particles like (a) from aqueous dispersions, even though this could be useful for nanoscale bioassays.

The ability to study the dynamics of self-assembly will be critical to developing this understanding. Electrophoretic deposition of nanoparticles will enable the nucleation, growth and melting of nanoparticle arrays to be driven using ac-electric fields. Small-angle X-ray scattering (SAXS) techniques can then assess the structural ordering length scale. With standard surfactant coatings, the spacing between nanoparticles in self-assembled arrays is large enough that magnetostatic interactions dominate exchange. By changing the particle size and spacing, purely magnetostatic ferromagnetism has been observed in these structures. The length scale of the structural ordering is also shown to be important; with a coherence length below _300 nm, the nanoparticle assemblies show spin-glass-like behaviour, while highly ordered structures act more like bulk ferromagnets.

An important research goal is to correlate the structural order and the nanoscale magnetization. SANS results show evidence of multiparticle magnetic correlations, but because large samples are needed, these experiments cannot be carried out on arrays with small numbers of layers. The unusual preference for arrays with an odd number of layers, provide information about the layer-by-layer magnetization pattern.

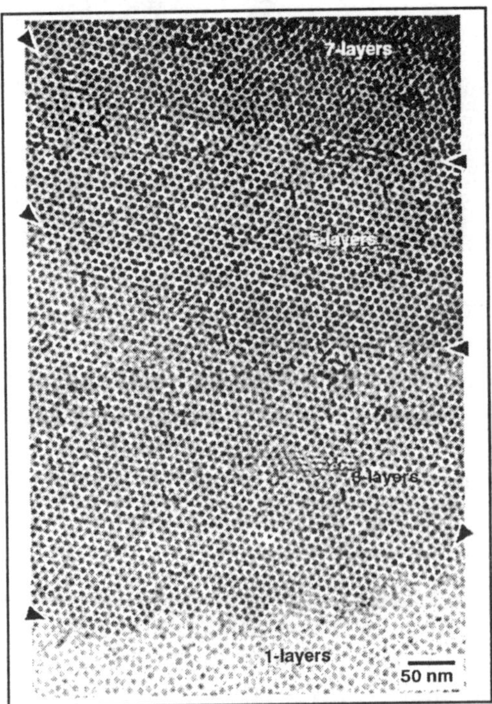

Fig. Arrays of 6 nm Fe Nanoparticles Showing an hcp Lattice and a Preference for an Odd Number of Layers.

Early resonant soft X-ray scattering studies have shown that it is possible to isolate pure magnetic scattering reflecting average interparticle moment orientations, and analyse these to understand the significance of dipolar interparticle interactions, even in superparamagnetic assemblies. The large resonant cross-sections enable X-ray techniques to study assemblies as thin as one particle layer. These capabilities will be useful in evaluating the degree of exchange coupling of nanoparticles, either to each other or to other magnetic structures in nanoscale devices.

Computational Foundation of Cluster Magnetism

Fundamental to understanding the magnetic behaviour in clusters is the evolution of the magnetism as the structural scale descends from the bulk to the nanoscale. It is well established from decades of research that interface magnetism can be quite complex and different from bulk magnetism. Due to reduced symmetry, the magnetic anisotropy can be orders of magnitude larger than in the bulk. This result can lead to magnetic frustration and reorientation of the magnetization at the surface and interface. Furthermore, interfaces between dissimilar materials can change their individual properties. For example, when in contact with an antiferromagnet, the properties of a ferromagnet change dramatically; the coercive field is enhanced and, the magnetization curve can become asymmetric showing the exchange bias effect.

In nanoscale magnetic clusters, interface effects are expected to be even more significant as the interface region is a dominant part of the entire structure. For example, in an FePt nanoparticle of 6-nm diameter, –40% of the atoms are located within one lattice constant of the surface. Indeed, surface segregation and surface magnetism are then expected to have a dominant effect on the magnetic state. Kodama et al. found that the increase in coercive field of $FeNi_2O_4$ nanoparticles over that of the bulk is due to a spin-glass-like surface spin structure to which the ferrimagnetic core of the particle couples. Understanding the complex atomic spin structure of magnetic nanostructures is thus essential to the mastering of nanomagnetism itself.

Measuring the non-collinear magnetic state of a nanostructure is a difficult task since most experimental probes of magnetism average over length scales of X10 nm. Information about the atomic-scale spin structure is usually inferred by comparison with model calculations. In many nanoscale systems, however, calculations cannot be based on bulk models, since parameters such as exchange and anisotropy constants, as well as the composition profile and atomic valence, are no longer similar to their bulk values. Model calculations, therefore, must be based on first-principles computational techniques that require input only of the atomic number, and are capable of predicting non-collinear magnetic states, atomic moments, dynamics and the response to external stimuli. Methods to calculate the electronic structure of magnetic nanosystems are mostly based on the local spindensity approximation (LSDA) to density functional theory (DFT).

These methods have proven to be reliable in predicting ground state properties, such as magnetic moments and anisotropies, as well as to calculate the temperature of magnetic phase transitions. Extensions of the LSDA to study the dynamics of non-collinear spin systems (in particular constrained LSDA) are now available. Such efforts have been successfully applied to predict non-collinear spin structures of complex antiferromagnets such as FeMn as well as the role of induced moments on high susceptibility elements such as Pt and Pd, or to calculate the effects of Ru on the magnetic exchange and anisotropy in FePt, CoPt, and FeRu, and reorientation transitions in thin films. Furthermore, the application of orbital dependent functionals have matured to the point where it is now possible to reliably treat from first principles strongly correlated magnetic systems, such as manganites and dilute magnetic semiconductors.

In recent years, the development of order-N electronic structure methods and their implementation on modern high performance computing hardware have made possible simulations of non-collinear spin systems with thousands of atoms in the unit cell. With these capabilities, it is possible to show, for example, that the spinstructure of FeMn in a Co/FeMn thin film heterostructure reorients from the bulk 3Q structure into a quasi-1Q structure in the thin film with preferential moment directions perpendicular to the ferromagnetic Co moments. Present-day calculations are already large enough to allow prediction of magnetic ground state properties of nanostructures.

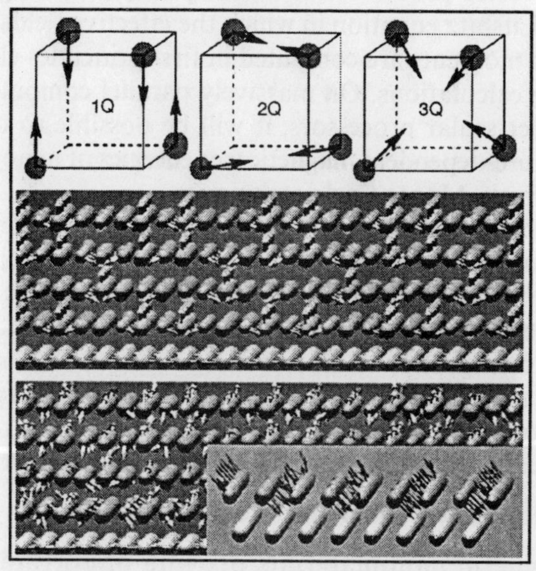

Fig. Upper Panel: Three Spin States Consistent with Neutron Scattering Measurements of d-FeMn.

Mo¨ ssbauer spectroscopy indicates that the 1Q state is not present, while first principles calculations show that the 2Q and 3Q states are the only stable states. The latter is energetically the most favorable. Middle panel: starting spins configuration for a first principles calculation of a FeMn/Co multilayer. Co (green) is assumed to be in a ferromagnetic state while FeMn is initiated in a perfect 3Q state. Lower panel: final configuration after full relaxation of the magnetic state. Co atoms are still perfectly ferromagnetically ordered. The spin structure of FeMn has rearranged and resembles a 1Q state with Fe and Mn moments aligned perpendicular to the Co moments.

It is expected that computer performance will increase by a factor of 100–1000 in the next five to ten years. With these advanced capabilities it will be possible to study finite temperature properties and dynamics of the magnetic state in nanostructures, for example nanoparticles of 5–10 nm diameter consisting of 5,000–20,000 atoms. On vector computers with 104 fast vector processors, it will be possible to integrate the Landau Lifshitz equation in which the effective fields and the magnetic moments are computed in first principles electronic structure calculations. On massively parallel computers with 4105 super scalar processors, it will be possible to calculate temperature-dependent magnetic free energies of nanoparticles with ab initio Monte Carlo techniques.

In these approaches, the energies of individual spin configurations are calculated with first principles electronic structure methods, and novel Monte Carlo techniques are used to sample configuration space, thus enabling calculation of the entropic contribution to the free energy. Due to the dominance of magnetic fluctuations at the atomic scale, incorporating temperature and dynamics effects in predictions of magnetic properties is crucial for understanding nanostructures.

Magnetic Clusters for Biomedical Applications

Magnetic nanomaterials provide opportunities for diagnostics and therapeutics for both in vivo and in vitro applications in biomedicine. The size scale of nanoparticles—similar to that of common biomolecules –makes them interesting for a wide range of biomedical applications that includes intracellular tagging, contrast agents, antibody targeting and hyperthermia protocols. For reference, size ranges of biological entities are given as follows: proteins (5–50 nm); genes (2nm–10–100 nm); viruses (20–450 nm) and cells (10–100 mm). The combination of biology and magnetism found in this arena is useful, because the biochemistry enables a selective binding of the nanoparticles, while the magnetism renders them easy to manipulate.

Moreover, the absence of ferromagnetism in most biological systems allows the signal from magnetic nanoparticles to be readily detected with low noise. Directed drug delivery can be accomplished with magnetic nanoparticles: a drug is bound to a magnetic particle and either DC-magnetic fields are used to confine the drug in a specific location of the body, or ac-magnetic fields are used in order to trigger the release of the drugs.

Key challenges for nanoparticles engineered for biological applications include the modification of nanoparticles for enhanced aqueous solubility, biocompatibility or bio-recognition, optimization of their magnetic properties including their relaxation dynamics over a broad range of frequencies and applied fields. For example, contrast in MRI is produced when magnetic nanoparticles modify the magnetic relaxation of the surrounding tissue. In addition to surface functionalization, nanoengineering of particle surfaces to optimize both their magnetic and optical response is important for diagnostic and therapeutic applications.

Hyperthermia involving the controlled heating of tissue employing magnetic nanoparticles to promote targeted cell destruction has the potential to be a powerful cancer treatment. Engineering superparamagnetic nanoparticles to yield the appropriate heat generation under AC magnetic field excitation is a complex process, highly dependent on particle size, crystallinity and shape. Investigation of magnetic heating parameters suitable for biological applications involves particle size and magnetic anisotropy optimization as well as studies of the complex magnetic susceptibility, heat capacity and power dissipation. Magnetic nanoparticles coated in a lipid bilayer, known as magnetoliposomes, can combine heat therapy with drug delivery to provide a synergistic treatment strategy. Magnetic particles can be injected into a patient and guided to a target site with an external magnetic field and/or specifically bind to target cells via recognition molecules coated onto the surface of the particles.

Optimum performance requires control over the synthesis of the nanoparticles and their surface modification. While biomcompatible iron oxide nanoparticles are the materials of choice for these applications, specific applications call for higher-moment cobalt nanoparticles. To this end, a synthesis route was recently developed involving the rapid decomposition of metallorganic precursors in a surfactant environment to prepare cobalt nanocrystals with tailored sizes (diameter –5–25nm) and controlled shapes, such as spheres or disks. A surfactant coats the particles during synthesis and plays a key role in the nucleation and growth of the particles and their geometry. For example, disk-shaped nanoparticles

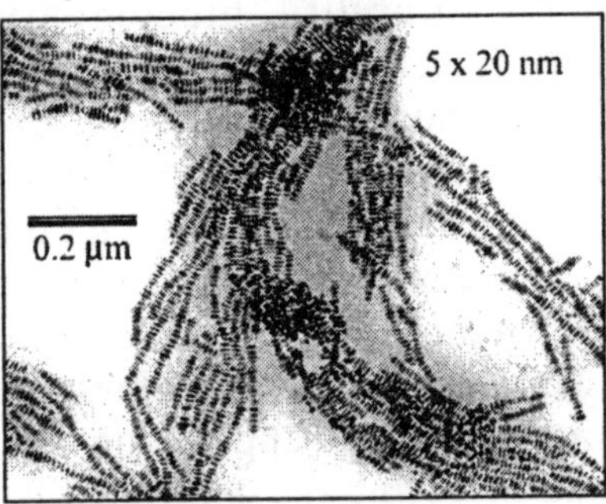

Fig. TEM Micrograph of Self-assembled Ccobalt Nanodisks: Lyotropic Liquid Crystals.

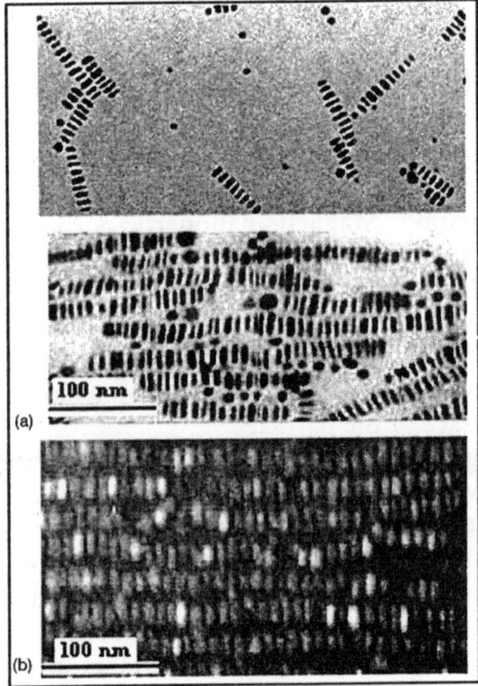

Fig. Enlarged Views of Cobalt Nanodisks.

are obtained by judicious choice of surfactants that adhere selectively to certain crystallographic planes promoting anisotropic particle shapes. While as-synthesized particles are initially hydrophobic, coating the particles with a second lipid layer (1,2-dipalmitoyl-sn-glycero-3-phosphocholine [DPPC]) is an effective route to make them hydrophilic and biocompatible.

The preparation of biologically functionalized magnetic nanoparticles can be quite challenging. A novel approach to the creation of appropriately sized particles is to utilize viruses as inert molds for the nucleation and growth of magnetic nanoparticles. A virus has a rigid container (the capsid), constructed from various proteins.

This capsid contains the DNA that the virus injects into a living cell to replicate itself. For some of these viruses it is possible to remove the DNA, leaving behind an intact and inert capsid shell (ghost virus). The empty capsid can subsequently be used as a template for fabricating magnetic nanoparticles via precipitation from solution. This synthesis strategy has been demonstrated for iron oxide and metallic cobalt nanoparticles using T7 bacteriophage, which produces nominally spherical particles of 40-nm diamete. The magnetic nanoparticles fabricated by this procedure ("magnetic viruses") have several key advantages to other magnetic nanoparticles fabricated using more traditional pathways. As all the viral particles are identical, they present a uniform template for the growth of the magnetic nanoparticles resulting in a very narrow size distribution.

Furthermore, the size and shape of the particles can be varied by using different virus templates. Magnetic viruses can be engineered to express the desired biological functionality, since phage display libraries (collections of many different viruses with slightly different DNA) can be utilized to select viruses with the required affinity for the target of choice. In addition, virus capsids are inherently biocompatible and thus ideally suited for in vivo applications.

The magnetism of small iron-based clusters may also lead to new diagnostic capabilities of neurodegenerative disorders, particularly Alzheimer's disease (AD), Parkinson's disease and Huntington's disease. It is known that patients with these diseases experience a disruption of normal iron metabolism, with total iron levels in diseased brains being significantly higher than in healthy ones. Cellular iron storage takes place in ferritins, a family of iron-storage proteins that sequester iron inside a protein coat as a hydrous ferric oxide–phosphate mineral similar in structure to the non-magnetic mineral ferrihydrite.

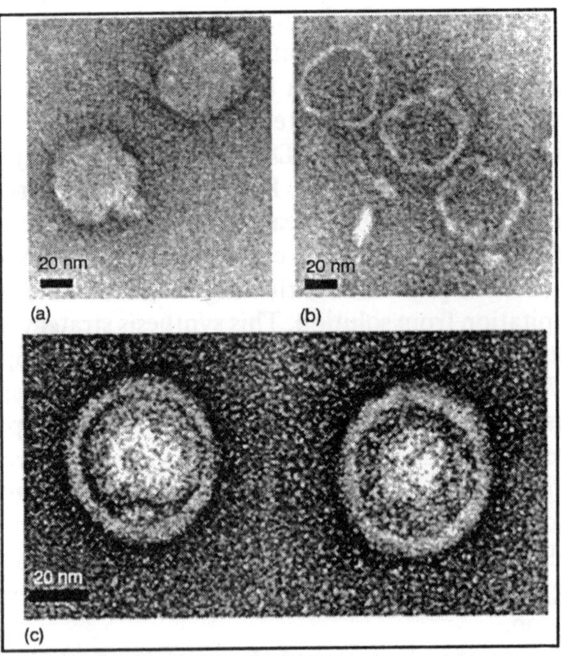

Fig. TEM Images of T7 Bacteriophage. (a) Normal Viruses, (b) Ghost Virus Particles After Osmotic Shock, and (c) "Magnetic Viruses" with Iron Oxide Nanoparticles at Their Centre

A likely explanation of the elevated iron levels in affected brains is due to increased iron loading of the ferritin cores. In normal ferritin, a rigid protein shell surrounds an 8-nm-diameter cavity partially filled with ferrihydrite, with extra room in the cavity to accommodate additional iron. Recent findings show that increased iron loading in brain tissue ferritin may lead to the formation of magnetite, a ferrimagnetic mineral that has Fe^{2+} ions, in contrast to the Fe^{3+} ion found in ferrihydrite.

This biomineralization transformation involves the reduction of Fe^{3+} to Fe^{2+} and may be accompanied by toxicity and the production of free radicals associated with the presence of Fe^{2+} ions, believed to underlie oxidative stress and neuron damage. Studies of the ferritin-to-magnetite transformation using a series of AD and control tissue samples indicate a striking correlation between the amount of magnetite present and the progression of the disease.

COMPLEX TRANSITION-METAL MAGNETIC SYSTEMS

Despite a long history of research starting soon after WWII, there is a resurgence of interest in multi-element transition-metal magnetic compounds that exhibit electronic and magnetic phase transitions driven by a variety of parameters such as magnetic field, pressure and temperature. This is largely due to the discovery of the colossal magnetoresistance (CMR) effect. CMR involves increased conduction electron mobility upon aligning the core spins in a magnetic field. The sensitivity of the magnetic transition to physical perturbations implies a strong coupling between charge, spin and lattice degrees of freedom, thus illustrating the sensitive nature of the d-shell electronic states to small changes in the atomic environment.

Prototypical examples of these and related materials include the perovskite-structure manganites and cobaltites, as well as high-Tc cuprate and ruthenate superconductors. In his seminal book, Magnetism and the Chemical Bond Goodenough quantified the connections between orbital occupancy and geometry and the resultant magnetic character in TM compounds, notably in TM oxides and chalcogenides. Since that time a profusion of important discoveries have been made concerning the rich variety of magnetic phenomena and transformations in this family of materials. Recently, the ability to synthesize new, nanostructured complex materials and probe their electronic states and dynamic response has deepened our understanding as well as opened fresh questions. In particular, new discoveries concerning the interplay of closely spaced energies—thermal, magnetic, electrostatic—reveal unexpected ground states. In some cases electronic disproportionation takes place on the nanoscale, leading to a multiphase state that has only a very subtle chemical signature, if at all.

Complex TM magnetic systems raise a wealth of fundamental questions and possess undeveloped technological potential. Clarification of the thermodynamic underpinnings of the ground state of the material is key to understanding its behaviour. Many complex, strongly correlated systems exhibit quantum phase transitions, i.e., they undergo a phase transition at zero temperature as some parameter is varied.

This zero-Kelvin state, is found in systems that exhibit critical phenomena, including critical points and critical end points of various sorts, higher-order critical points, and entities coexisting in distinct phases. In such systems susceptibilities diverge despite the fact that only short-range correlations exist. Furthermore, the low-energy properties of the system are dominated by extremely rare configurations. Questions of theoretical interest include whether this phase separation constitutes a true thermodynamic state, and how the phase transition percolates throughout the system. Additionally, it is important to determine to what extent scaling works, and whether the phase separation has temporal characteristics of interest. In addition to quantum phase transitions, complex systems have recently been confirmed to exhibit subtle phase coexistence that is thought to underlie their phase and electronic transport properties.

Basic questions about these states persist, such as what is the size of the phase regions, and how does one control the size and morphology? The extent of local chemical variation remains unknown, as does the role of strain, disorder and lattice defects. It is of profound importance to identify structural precursors to phase separation in this family of materials.

Multiple degrees of freedom in complex systems can give

rise to multiferroic phenomena, whereby a material may exhibit simultaneous ferromagnetism, ferroelectricity and ferroelasticity, subjected to specific crystal symmetry constraints.

The functionality of this class of materials extends beyond that of spintronics since they possess the potential of their parent form as well as crossover effects, permitting an additional degree of freedom—multifunctionality—in device design. Possible devices employing these novel materials include multiplestate magnetic elements, electric-field-controlled magnetic resonance devices, and transducers with magnetically modulated piezoelectricity. Furthermore, there is great technological interest in complex TM magnetic systems.

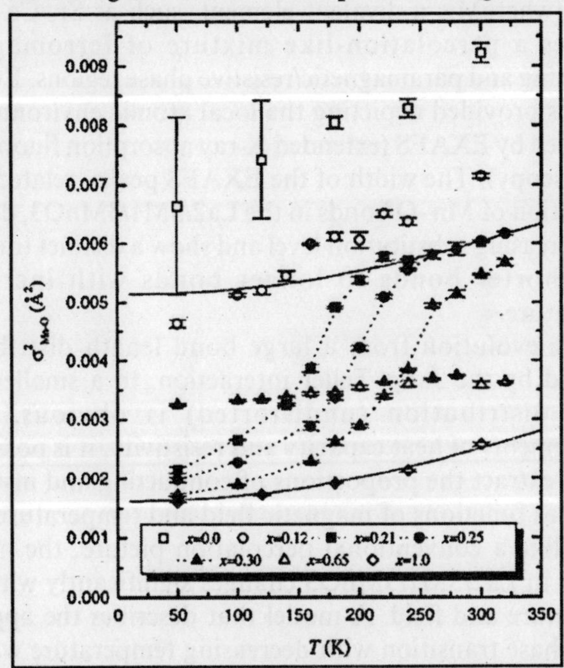

Fig. Width of EXAFS Peaks in La1–xCaxMnO3 for Various Concentrations x. The Increase in Width Reflects the Formation of Polarons at High Temperature. Dotted Lines are Guides to the eye. Relative errors are Estimated to be Smaller than the Symbols, Except for LaMnO3.

This is due to the collection of extraordinary functional phenomena that occur in the vicinity of the magnetostructural phase change, which are manifestations of the highly-correlated electron behaviour. It may be that material inhomogeneity underlies these phenomena. In oxides, these phenomena include the CMR effect found in the holedoped (La,Sr)-based manganites, of interest for magnetic sensors and the giant magnetocaloric effect (MCE) that is under development for CFC-free magnetic refrigeration.

Fundamental Properties: Phase Separation and Griffiths Phase in Doped Manganites

The CMR of the prototypical compound La2/3M1/3 MnO_3 (whereMis a divalent element, such as Sr, Ca or Pb) involves a percolation-like mixture of ferromagnetic/conducting and paramagnetic/resistive phase regions. Evidence of this is provided depicting the local atomic environment as quantified by EXAFS (extended X-ray absorption fluorescence spectroscopy). The width of the EXAFS peaks, related to the distribution of Mn–O bonds in the La2/3M1/3MnO3, decrease with increasing substitution level and show a distinct transition from shorter bonds to longer bonds with increasing temperature.

The evolution from a large bond length distribution, distorted by the Jahn–Teller interaction, to a smaller bond length distribution (undistorted) is obvious. From measurements of heat capacity and resistivity, it is possible to directly extract the proportions of conducting and insulating regions as functions of magnetic field and temperature.

Unlike a conventional percolation picture, the metallic fraction in La2/3M1/3MnO3 changes significantly with both temperature and field. A model that describes the approach to the phase transition with decreasing temperature was first presented by Griffiths, and leads to what is termed a Griffiths singularity. Within this context, at any temperature, pre-existing ordered regions are coalesced into clusters. As the temperature is lowered, the clusters grow in number but not in size. This results in a sudden connection of clusters, leading to percolation—but of a very special sort. The Griffiths phase and nanoscopic phase separation is discussed by Burger et al.

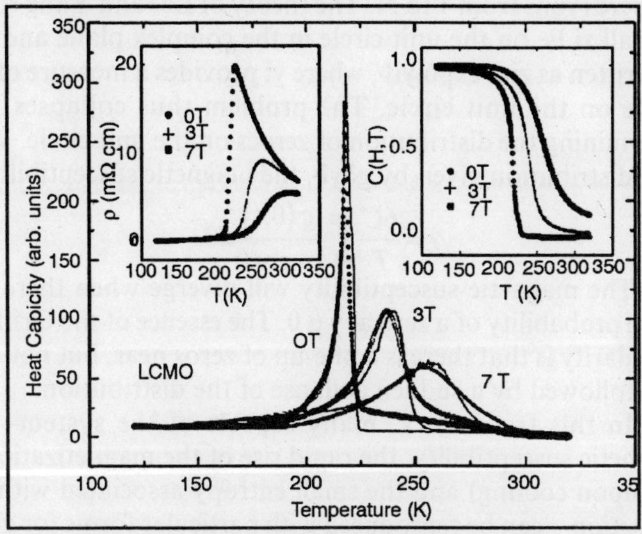

Fig. Heat Capacity as a Function of Field for La0.67Ca0.33MnO3 and a Calculation Based on a Two-state Model. The Insets Show the Resistivity and the Metallic Fraction c(H) Extracted Using an Effective Medium Approach.

The Griffiths phase encompasses that range of temperatures between the maximum possible transition temperature TG and the observed one at the Curie temperature TC.. The treatment of the Griffiths phase was extended by Bray and is based on the Lee and Yang approach to phase transitions. In this framework the partition function Z for a system undergoing a phase transition can be expressed as a polynomial with all positive terms that may be factored into N complex roots, where N is the number of particles in the system.

$$Z = e^{Nh}\prod_i (z - z_i)$$

and

$$\frac{F}{kT} = Nh + \sum_i (z - z_i),$$

where in both expressions z µ expð_2 hÞ, h is the scaled magnetic field and the product or sum is taken over the index i, where i runs from 1 to N. The theory of Lee and Yang states that all zi lie on the unit circle in the complex plane and can be written as zi µ expðiyiÞ, where yi provides a measure of the angle on the unit circle. The problem thus collapses into determining the distribution of zeroes on the unit circle. With that distribution given by gðyÞ, the magnetic susceptibility is

$$\chi = \frac{6C}{T}\int_0^{\pi} \frac{g(\theta,t)}{1-\cos\theta}\,d\theta.$$

The magnetic susceptibility will diverge when there is a finite probability of a zero at y µ 0. The essence of the Griffiths singularity is that there is a pile-up of zeros near, but not at, y µ 0, followed by a sudden collapse of the distribution.

In this framework, many aspects of the system—the magnetic susceptibility, the rapid rise of the magnetization at Tc (upon cooling) and the small entropy associated with the transition—can be reproduced with particular forms for gðyÞ. Furthermore, the lower the value of Tc is, the less entropy change there is in the vicinity of the transition. The large number of Yang–Lee zeroes at small y translates into a large number of regions with large susceptibility. This makes the effective spin large in the Griffiths phase, and releases much of the ordering entropy well above the transition. To the extent that the low temperature phases are suppressed by disorder, the region between the "pure" transition and the actual one may be indeed a Griffiths phase. Hence, an interesting direction for high resolution magnetic scattering is to follow the development of these numerous, highly correlated magnetic regions and to study their evolution with magnetic field and temperature.

Inhomogeneity in Complex Magnetic Systems

Inhomogeneities can be driven thermodynamically by chemical doping or competing interactions at the nanoscale, as is the case of CMR and high-Tc superconductors, or introduced via artificially tailored synthesis, as is the case of magnetic nanocomposites. Inhomogeneity can also be thought of as a self-assembly process driven by quantum fluctuations (at least for those systems sitting near quantum critical points or with nearly degenerate phases.) The existence of inhomogeneity in complex systems has been conclusively proven to exist experimentally in many systems and this inhomogeneity can indeed be reproduced by phenomenological models. Examples of this may be found in CMR materials, in which ferromagnetic metallic regions coexist with antiferromagnetic insulating regions at the nanoscale. In high-Tc superconductors, nanoscale inhomogeneities in all electronic, magnetic and structural degrees of freedom have been observed. Inhomogeneities exist by design in magnetic nanocomposites, such as in exchange-spring magnets that consist of hard and soft magnetic phases mixed at the nanoscale in a quest to achieve improved permanent magnetic properties. Furthermore, inhomogeneities may also arise at interfaces from the complex interplay between the electronic structures of dissimilar materials.

The existence of nanoscale ferromagnetic clusters in many materials previously thought to be homogeneous is now without question. It is thus necessary to inquire about the origins of the effect. It is not clear if the phase transition is one of nucleation and growth (thermodynamically first-order) or, as in spinodal decomposition, fundamentally second-order. Open questions under study currently include the influence of temperature and field on the phase separation, and whether the phase separation is reversible. Importantly, what are the structural precursors responsible for the phase separation phenomenon, if any?

These need not be gross chemical segregation, but could be far more subtle, as e.g. in the LSMO manganites CMR material, spatial inhomogeneities in the oxygen stoichiometry, valence, Mn–O bond length, or very short length scale variations in the La/Sr ratio even in a nominally randomly doped system. The challenge is to determine if phase-separation phenomena are merely pinned by, as opposed to due to, the inevitable inhomogeneity of dopant distribution in such materials. In this regard, one critical issue concerns the question of the length scale of the phase components of a multi-phase complex system. In some systems segregation occurs on nanoscopic scales, while in other systems micron-sized segregation occurs. It is possible that one case originates from intrinsic magnetoelectronic segregation, while the other is due to structural inhomogeneity.

Such questions naturally lead from identification of phenomena to their exploitation. For example, understanding the segregation mechanism in ferromagnetic phase-separated materials may lead to the creation of novel phase morphologies.

Exciting possibilities resulting from control over the phase separation include self-assembly of 3D magnetic matrices consisting of lamellar, gyroidal or cylindrical (filamentary) geometries. Such architectures are anticipated to have technological implications, for example in GMR superlattices and exchange bias-based devices.

It is becoming increasingly clear that the underpinnings of inhomogeneity, such as local defects and chemical disorder, remain essential components of the physics of many material systems of interest (from heavy fermions to manganites to ferromagnetic semiconductors). The change in magnetization of the ferromagnetic semiconductor (Ga, Mn)As with annealing at low temperatures (250 1C) for relatively short time periods from 10 min to 24 h. The considerable changes in the form of the magnetization curve, as well as in the Curie temperature, can be attributed to the important role of defects.

The atomic viewpoint of heterogeneity is extended to include site-specificity, in addition to element-specificity. Such studies provide understanding of the effects of inequivalent atomic environments on magnetic properties.

For example, CMR and high-Tc oxide materials, as well as intermetallic permanent magnet compounds Nd2Fe14B and Sm2Co7, all possess inequivalent crystal sites. Examples of these inequivalent crystal sites are the Mn3+ and Mn4+ sites in the CMR manganites and the the Cu(1) and Cu(2) sites in the CuO2 planes of the cuprate superconductors.

There are two Nd sites in Nd2Fe14B, (the 4f and 4g sites,

in Wykcoff notation) that are recognized to control the anisotropy of the compound.

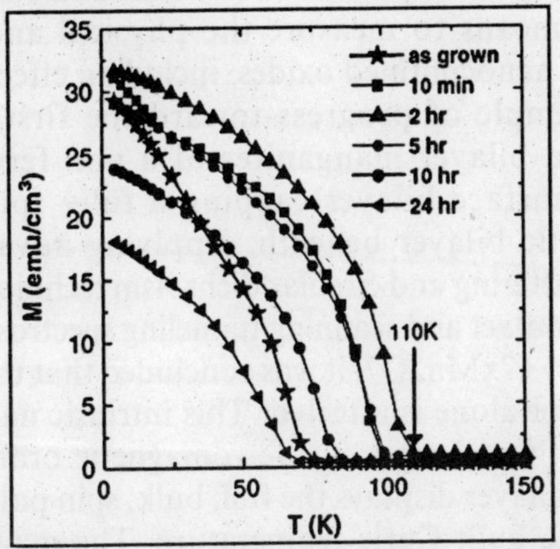

Fig. Change in magnetization of (Ga,Mn)As with annealing at 250 1C for the specified time periods. The changes in the magnetization and Curie temperature are attributed to the important role of defects, in particular the out-diffusion of Mn interstitials which are induced by the low temperature MBE growth required to make this material.

Techniques such as magnetic resonant X-ray diffraction using circularly polarized X-rays can separate the magnetic contributions of these inequivalent sites. Selection of a scattering vector along the high-symmetry direction in the tetragonal P42=mnm Nd2Fe14B structure allows structure factor contributions to the scattering from either one of the the other Nd site to nearly vanish. This feature enables the determination of the magnetic behaviour of each Nd site to be determined independently. In addition to atomic defects, inhomogeneity at the micrometer scale such as grain structure may play an important role.

For example, complex oxides are so sensitive to stoichiometry and strain that surfaces can exhibit significantly altered properties from the bulk. For nanostructures exhibiting 3D confinement, surfaces can have a dominant effect. However, the surfaces of bulk oxides are not completely understood. Thus, understanding the role of surfaces in confined oxides has three aspects: (i) the need to develop a fundamental understanding of unconfined surfaces; (ii) the need to clearly distinguish between bulk and surface effects; and (iii) a means to measure the physical and electronic structure of nanoconfined oxides, including effects of strain.

An example of progress toward the first goal is the discovery in bilayer manganites of a non-ferromagnetic, insulating surface bilayer capping a fully spin-polarized ferromagnetic bilayer beneath. Applying advanced X-ray magnetic scattering and circular dichroism techniques together with point-contact and scanning tunneling spectroscopy (STM) to La2_2xSr1+2xMn2O7, it was concluded that the outermost Mn–O bilayer alone is affected. This intrinsic nanoskin is an insulator with no long-range ferromagnetic order, while the subsequent bilayer displays the full, bulk, spin polarization up to nearly the bulk Curie temperature. The magnetic 'bulk' bilayers are expected to be metallic due to hopping between Mn+3 and Mn+4 ions by double exchange in this doped manganite, while tunneling data suggest that the surface bilayer is an insulator. The abrupt changes in only the topmost bilayer are likely due to the weak electronic and magnetic coupling between bilayers engendered by the crystal structure.

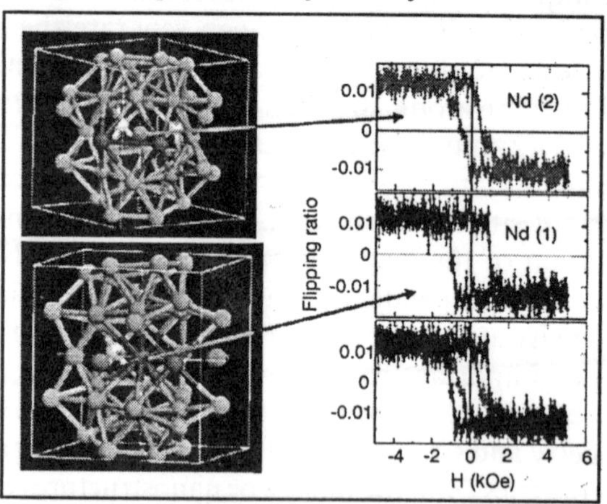

Fig. Coexisting Nd Crystalline Environments in Nd2Fe14B Permanent Magnet (left). Site-specific Nd Hysteresis Loops (right). The Crystalline Environment at the Nd(1) Site Results in its Higher Stability Against Demagnetizing Fields.

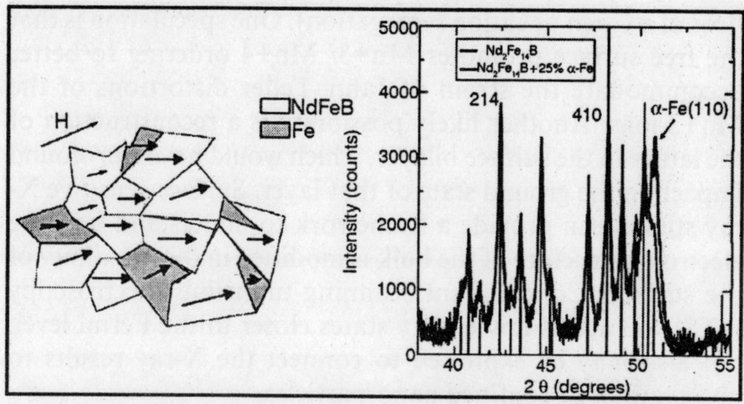

Fig. Left: Schematic of exchange-coupled nanocomposite with hard and soft magnetic phases. Magnetic contributions from Fe atoms in either phase can be separated by selecting appropriate Bragg diffraction conditions (right) and using circularly polarized X-rays at Fe resonance to couple to their magnetic moments.

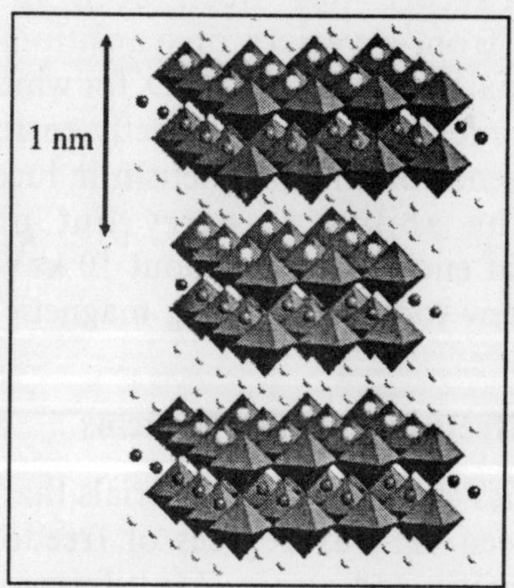

Fig. Structure of the naturally bilayered manganite, La2–2xSr1+2x Mn2O7. The MnO6 octahedra denoted in blue and (La,Sr) sites shown as yellow and red dots. The bilayer repeat distance is _10A Ï. The nonferromagnetic surface bilayer is colored brown.

The reason that the surface layer remains insulating in the layered manganite, above, is unclear. Surface sensitive spectroscopic measurements seem to exclude chemical disorder (loss of oxygen or cation segregation). One speculation is that the free surface promotes Mn+3/ Mn+4 ordering to better accommodate the strain of Jahn–Teller distortions of the Mn+3 ions. Another likely possibility is a reconstruction of the lattice in the surface bilayer, which would have a profound impact on the ground state of that layer. Surface-sensitive X-ray studies can provide a framework to understand how the electronic structure of the bulk is modified in the proximity of the surface. Concomitant scanning-tunneling spectroscopy (STS) will access low energy states closer to the Fermi level, but also may be exploited to connect the X-ray results to observations in confined nano-particles.

As another recent example related to the third goal above, it has recently been observed that more bulk sensitive hard X-ray photoemission at _6 keV photon energy exhibits additional structure in Mn 2p core spectra from LSMO manganites that can be directly connected with doping-induced states. Similar additional structure has been seen in high-energy photoemission from a variety of transition-metal oxides, including epitaxial thin films of LSMO, for which the presence/ absence of these features can be directly correlated with the presence or absence of ferromagnetism induced by strain in the layers. The ability to carry out photoemission measurements at energies up to about 10 keV thus provides an important new tool for studying magnetic materials and nanostructures.

Multiferroic Effects in Complex Systems

Multiferroics are intriguing materials that exhibit strong coupling between various degrees of freedom: structural, electrical, magnetic and strain. Multiferroic materials are single-component materials or composites exhibiting two or more ferroic features such as ferromagnetism, ferroelectricity, or ferroelasticity/shape-memory effects. While there are a number of materials that possess ferroic properties (i.e., ferromagnetism and ferroelasticity), there need not be a large coupling between the two properties. Hence multiferroics, by their very nature, are identified as complex materials. The

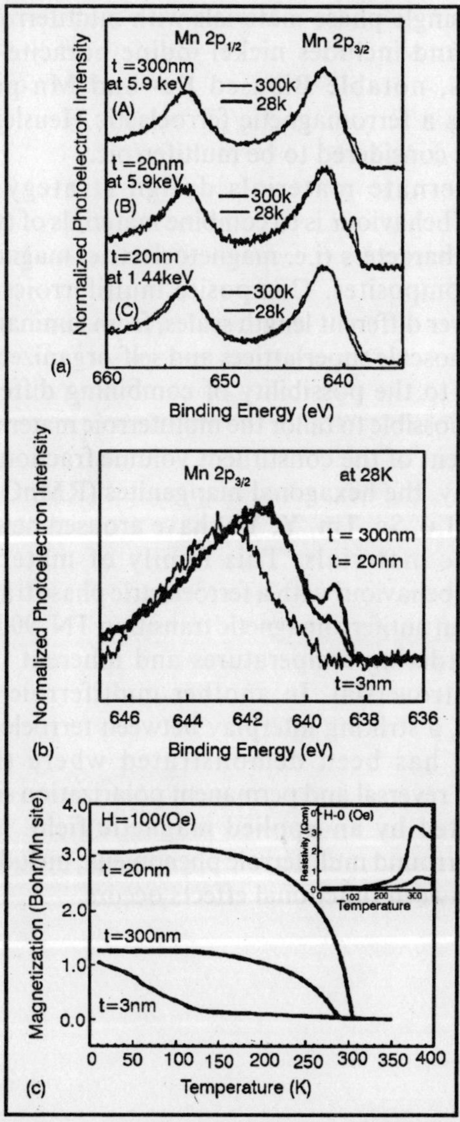

Fig. (a) Mn 2p spectra from (La0.85Ba0.15)MnO3 thin films, illustrating the additional feature at 639 eV binding energy observed only with more bulk sensitive 6 keV excitation. (b) The thickness dependence of this extra feature. (c) Measurements of magnetization M and resistivity r as a function of temperature, illustrating the correlation between them and the strength of the additional feature.

number of single-phase materials with multiferroic potential is limited, and includes nickel iodine boracite and mixed perovskites, notable Bibased Fe- and Mn-perovskites. Ni_2MnGa is a ferromagnetic ferroelastic Heusler alloy that may also be considered to be multiferroic.

An alternate materials design strategy to obtain multiferroic behaviour is to combine materials of two different functional characters (i.e. magnetoelectric, magnetostrictive) to create composites. Composite multiferroics have been produced over different length scales, from laminar geometries through nanoscale superlattices and self-organized structures. In addition to the possibility of combining different ferroic effects, it is possible to tailor the multiferroic materials response via adjustment of the constituent volume fractions.

Recently, the hexagonal manganites (RMnO3, with R µ Er, Ho, In, Lu, Sc, Tm, Y, Yb) have aroused new interest as multiferroic materials. This family of materials shows multiferroic behaviour with a ferroelectric phase transition near _900K and an antiferromagnetic transition TN_90 K, although the exact ordering temperatures and inherent interactions remain controversial. In another multiferroic compound ($TbMn_2O_5$), a striking interplay between ferroelectricity and magnetism has been demonstrated where the electric polarization reversal and permanent polarization signature are both actuated by an applied magnetic field. Many open questions surround multiferroic phenomena, including how the coupling between functional effects occurs.

Chapter 3

Synchrotron X-Ray Techniques and Relationship to Nanomagnetism Research

The application of synchrotron radiation to the study of magnetic materials has grown rapidly in recent years, owing in part to the availability of high-brightness synchrotron sources. Several characteristics of synchrotron radiation make the study of magnetic materials very attractive. First of all, the high brightness of the beam typically results in a flux of 1012_13 photons/s in less than a 1mm2 area, which enables the study of small or highly diluted samples.

The high scattering wavevector resolution due to the X-ray collimation and monochromaticity permits precise determination of magnetic modulations. Furthermore, the well-defined polarization characteristics of synchrotron radiation (naturally linear in the plane of the storage ring, but also variable in polarization via specially designed undulators), together with its relatively simple manipulation and analysis by crystal optics, can be used to study a variety of magnetization states.

Lastly, by tuning the energy of the incident beam near absorption edges (or resonances) of constituent elements, one can study the magnetic contributions of individual components in heterogeneous structures. Being able to vary photon energy over the approximate interval of 0.5–15.0 keV can also permit clearly distinguishing surface, interface, and bulk effects, as well as uncovering new effects that can be directly related to magnetism. X-rays interact with matter by scattering from both the electron charge and its magnetic moment, as well as by being absorbed via electron excitation and the photoelectric effect. The charge scattering is the dominant term and is the basis for most condensed matter studies using X-rays. Although small, the scattering from the magnetic moment is sufficient to extract valuable information on magnetic structures in single crystals.

Enhanced sensitivity to magnetic moments can be achieved by tuning the X-ray energy to selected resonances. These resonant enhancements have resulted in widespread applications of X-rays in the study of magnetism, both in the absorption (X-ray magnetic circular dichroism or XMCD) and scattering (X-ray resonant magnetic scattering) channels. These techniques include studies of interfacial magnetic roughness in multilayers, and of morphology of magnetic domains in buried interfaces. Furthermore, by performing polarization analysis of the scattered radiation or applying sum rules to dichroic absorption spectra of spin-orbit split absorption edges it is possible to distinguish between spin and orbital contributions to the magnetic moment in an element-specific manner. This is a unique attribute of magnetic scattering and spectroscopy techniques, and it is the primary reason why these techniques are powerful tools in magnetism studies.

Most synchrotron studies of nanomagnetism to date have been performed using soft X-rays (loosely defined as possessing energyo3 keV) since resonant dipolar transitions in this energy regime access electronic states carrying large magnetic moments in most materials (e.g., 2p-3d states in transition metals, and 4d-4f states in rareearths.) Hence, the magnetic signals are larger and easier to observe. Harder X-ray energies (4 3 keV) access electronic states with smaller, yet significant, magnetic moments (e.g., 1 s-4p states in transition metals and 2p-5d states in rare-earths). While experimentally more challenging, hard X-ray studies of magnetism offer unique advantages. The higher penetrating power of these X-rays enables the study of buried structures and interfaces, which can be important in characterizing a wide variety of magnetic systems used in modern technologies, such as permanent magnetic materials and artificial, thin-film heterostructures. The penetrating power of hard X-rays yields a true bulk-measurement probe without the need for high-vacuum conditions, while soft X-ray measurements are surface sensitive and must be performed in ultrahigh vacuum conditions. Furthermore, the shorter hard X-ray wavelengths permit diffraction studies to probe the magnetic order in both crystals and artificial, periodic nanostructures, such as multilayers and patterned dot/hole arrays.

The rich polarization dependence of magnetic scattering can be used to extract the magnetic ordering of a material. Antiferromagnetic (AF) structures are commonly studied with linearly polarized radiation. In the absorption channel, the linear magnetic dichroism (MLD) effect results in absorption contrast for parallel and perpendicular alignments of the X-ray's linear polarization and the sample's magnetization in the presence of magneto-crystalline anisotropy, which can be used, e.g., to image AF domains in exchange-biased systems. In the diffraction channel, AF ordering results in Bragg diffraction at the magnetic ordering's wavevector. As discussed below, synchrotron radiation brightness, together with resonant enhancement of the magnetic scattering cross section, can result in the detection of X-ray magnetic scattering from AF systems. Circularly polarized (CP) radiation can also be useful in studies of AF materials. CP X-rays were used for real-space imaging of chiral domains by helicity-dependent Bragg scattering from the spiral AF state of a Ho crystal by Lang et al.

Ferri- or ferromagnetic structures are commonly studied with CP radiation. In the absorption channel, XMCD results in absorption contrast for parallel and antiparallel alignment of the X-ray helicity and the sample magnetization. 216 By measuring this absorption contrast in spinorbit split core levels (e.g., transition metal L2 and L3 edges), element-specific magnetic moments in the final state of the absorption process (both spin and orbital components) can be extracted by application of sum rules. This contrast, in combination with focused X-ray beams, can be used to image ferromagnetic domains in nanostructures. The pulsed nature of synchrotron sources offers the possibility of studying time-dependent phenomena.

One of the desirable experimental capabilities is to achieve a high spatial resolution (_5 nm) and combine it with fast temporal resolution (_1 ps.) Neither of these capabilities are currently available at the APS. Time resolution depends on the electron bunch structure and, at the APS, the most common operating mode with 24 bunches equally distributed around the storage ring enables 153-ps time resolution. There is a considerable effort under way to reduce the time resolution to _1ps. Pump–probe type experiments, open new approaches to study the dynamics of magnetization reversal.

Selected examples will now be used to highlight the relationship of synchrotron-based X-ray techniques, both existing and under development, to address key scientific and technological issues associated with confined magnets. Then the complementarity of X-ray and neutron techniques are discussed, and the requirements for X-ray techniques and associated equipment developments in the future are outlined.

BURIED SOLID–SOLID INTERFACES

Buried solid–solid interfaces are ubiquitous in nanomagnetic structures. Critical to the characterization process is a detailed understanding of their chemical concentration profiles, element-specific chemical states and magnetic properties, as well as quantitative chemical and magnetic roughness parameters. For example, X-ray standing waves in the 0.05–10 keV range can provide powerful methods for studying such interfaces. These standing waves can be created by Bragg reflection from any multilayer structure, or from any set of atomic planes in a single crystal or an epitaxial sample. Interfaces that exhibit lateral spatial variations, or wedge profiles, permit the standing wave to be scanned through the interface simply by moving a focused X-ray beam along the thickness gradient.

In this example, a current signal that creates a transient magnetic field is applied synchronously with the X-ray pulse in order to study magnetization dynamics in permalloy dots grown on top of a coplanar waveguide. For most of the time, the APS storage ring runs in a 24-bunch operating mode delivering X-ray pulses separated by 153 ns and lasting _100 ps. In a 324-bunch special mode, the separation between X-ray pulses is 11.2 ns.

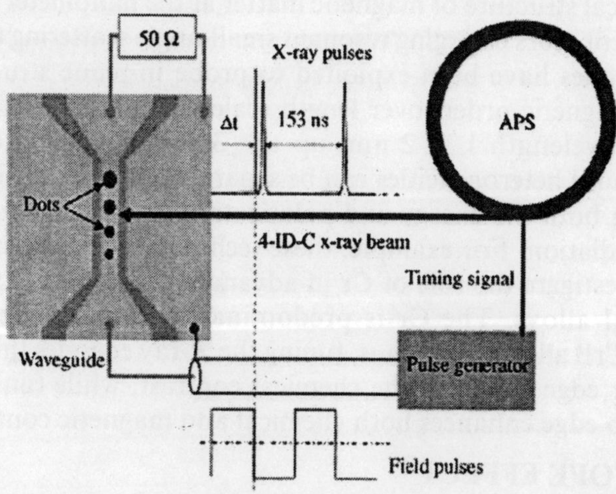

Fig. Schematic of a Typical Pump–probe Experiment at the APS.

Such data can be obtained under both resonant and non-resonant conditions, and with either photoelectrons or fluorescent X-rays as the detection mode. For instance, measurements for an Fe/Cr interface with an excitation energy _800 eV using, photoelectron detection yield the compositional variation of the buried interface, and via XMCD, the magnetization profile across the interface of Fe and Cr. The data obtained in this study, together with the concentration and magnetization profiles determined from it. Using even higher energies (5–10 keV range) would provide greater probing depths for both photoelectrons and X-rays, as well as stronger standing wave effects due to higher reflectivity. With a focused X-ray beam, such measurements would also provide lateral microscopy information. The status of various types of standing wave experiments is discussed elsewhere. The use of these techniques for studies of buried interfaces should provide solutions to many interesting and challenging problems in the field of nanomagnetism.

RESONANT SPECTROSCOPIES

Resonant spectroscopies of 3d TM films at the L edges provide various opportunities to investigate magnetic and chemical structure of magnetic matter at the nanometer scale. In recent years emerging resonant small-angle scattering (SAS) techniques have been exploited to probe in-plane structural and magnetic order, over length scales ranging from the X-ray wavelength l, 1–2 nm, up to _300 nm Magnetic and structural heterogeneities can be separated and quantified by tuning both the energy and polarization of the incoming X-ray radiation. For example, these techniques have been used to investigate the role of Cr in advanced magnetic recording medial alloys. The Cr is predominantly non-magnetic in CoPtCrB alloy media, thus, tuning the X-ray energies through the Cr edge enhances the chemical contrast, while tuning to the Co edge enhances both chemical and magnetic contrast.

ISOTOPE EFFECT

X-ray techniques can be element specific; however they normally do not distinguish between different isotopes. In contrast, neutron scattering is very sensitive to the specific isotopic composition. This strong isotope contrast can be used in the case of neutron scattering to enhance scattering contrast for magnetic structures by reducing the contrast for chemical structures.

However, X-ray techniques also become isotope-sensitive in the case of nuclear resonant scattering, which is related to the well-established Mo¨ ssbauer effect. This technique is restricted to Mo¨ ssbauer- active isotopes of elements such as Fe, Sn, Sm, Dy and Eu. Nuclear resonant scattering permits the measurement of nuclear hyperfine fields, which can be directly related to the magnetization vector. Several key differences from the more traditional X-ray techniques (such as XMCD) are: (i) high spatial selectivify can be obtained by selective preparation of the sample with Mo¨ ssbauer-active isotopes; (ii) the full direction of the magnetic moment in all three dimensions can be obtained; (iii) even for TM (such as Fe) the penetration depth is up to 100 nm, since hard X-rays can be used for the magnetic contrast (instead of soft X-rays at the L-edge); and (iv) they can be sensitive to atomic scale magnetic order, such as antiferromagnetism.

Complementarity of Neutron and X-Ray Techniques

Since there are two major types of instruments that are relevant to address some of the issues outlined above, it may be worthwhile to highlight advantages and drawbacks of each.

In the field of nanomagnetism, neutron scattering has several notable advantages, including the fact that it is a mature technique, it has substantial depth penetration, it is sensitive to isotopic substitutions, and the data interpretation is straightforward.

Its major disadvantage is low intensity and the necessity of large sample sizes to obtain data. Synchrotrons, however, provide high photon flux intensities (although this has to be normalized by the sensitivity to magnetism) and are element sensitive, but being relatively new, data interpretation is not necessarily direct. Moreover, the high beam intensity can make sample heating a problem.

Scanning the X-ray spot along the wedge permits scanning the standing wave through the Fe/Cr interface. Lower panel: intensity and magnetic circular dichroism results, together with X-ray optical theory permitting the derivation of the concentration and magnetization profiles shown.

An example of X-ray neutron complementarity appears in the recent work for Roy et al., on the depth profile of uncompensated spins in an exchange bias system. Synchrotron sources have been built to optimize for either scattering or spectroscopy experiments. However, recent advances in synchrotron instrumentation have tended to blur this distinction so that advanced scattering experiments can often be performed at the lower-energy machines, while many low-energy spectroscopy techniques can be implemented at the higher-energy machines.

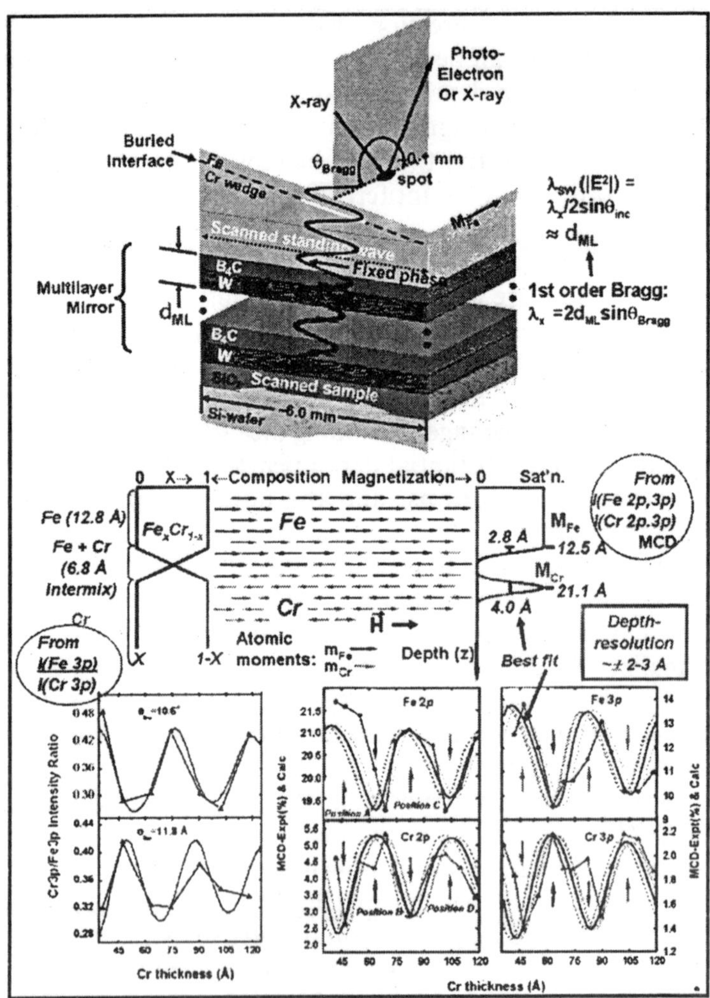

Fig.Upper panel: schematic view of an experimental configuration permitting the selective excitation of photoelectrons or fluorescent X-rays from a multilayer magnetic system.

X-RAY INSTRUMENT REQUIREMENTS

A key experimental feature for most applications is the quality and versatility of the sample environment. As first priorities it is crucial to be able to cover a diverse parameter space that captures the novel properties of the multitude of systems of interest. This includes:

- Temperature range: 1–500 K.
- Magnetic fields up to 12–20 T.
- 3D Magnetic field orientation.
- Ultra-low remanent field.
- Pressure range to include that available via diamondanvil cell techniques.
- Optical, electrical access.

Beyond these needs, of course, it would be useful to have ultra-low temperatures o1K and very high fields 420 T for select applications. Additionally, in situ sample preparation is of high priority in many experiments, especially for surface magnetism experiments. The APS can already offer a subset of choices from the menu above, such as a temperature range from 6–500K and magnetic fields up to 4T. Presently, there is a strong emphasis on the development of high-pressure capabilities up to 20 GPa via diamond-anvil cells in a magnetic field up to 1 T. In addition, plans are underway to introduce in situ transport and optical measurements in combination with X-ray techniques. Also, there are near-term plans to offer the scientific community access to high field capabilities up to 13 T. In the context of a synchrotron facility, effective implementation and efficient operation of most of these specialized instruments require either dedicated beamlines or dedicated end-stations. Since it is unrealistic to pursue parallel development of all new capabilities, the input of the nanomagnetism community will continue to be crucial in setting the proper priorities so that hard X-ray scattering, spectroscopy and microscopy techniques will remain indispensable tools in nanomagnetism studies well into the 21st century.

Medium and hard X-ray spectroscopies, applied with dichroism and spin resolution, as well as spatial and time resolution, should be key elements of the APS 10-year plan. Absorption and scattering will clearly be a part of this, as one of the standard hard-X-ray experiments, but it should be stressed that developing world-class facilities for photoemission, such as described in the next subsection, if possible, coupled with soft and/or hard X-ray emission/ absorption in the same chamber would also be extremely valuable.

SPECIFIC EXAMPLES OF DESIRABLE CAPABILITIES

Hard X-ray Photoemission

High-energy photoemission/diffraction/microscopy in the 5–10 keV range would be desirable:

Photoemission is one of the most powerful spectroscopic tools in synchrotron radiation science, but at the excitation energies of 20–1500 eV typically used today, it is a mixed surface and bulk probe, with electron inelastic attenuation lengths in the 0.5–2nm range. Exciting photoelectrons in the 5–10 keV range would provide enhanced bulk sensitivity, which could be varied via decreased takeoff angle. This advance would reduce sample transfer and preparation problems as well as permit the measurement of properties that are more truly representative of bulk behaviour well below the surface, as discussed in a prior section.

The fact that bulk-associated spectral features are seen in hard X-ray photoemission that do not appear in soft X-ray photoemission also points out that we cannot hope to fully understand electronic and magnetic structure in complex systems without making measurements in both regimes of energy. In this regard, overall instrumental resolutions of about 50 meV have already been reported.

Such an experimental system should be mated to a variable polarization undulator with XMCD/MLD capabilities and a finely-focussed X-ray beam; the APS is planning such a beamline that will go to 50 nm. Additional capabilities would permit measuring element-specific electronic structure, magnetic spin and orbital moments, and valence-level densities of states, all with spin resolution.

One recent example of the power of applying core level and valence band spectroscopies to magnetism is found in studies of the CMR oxide LSMO carried out in the soft X-ray regime.

In this study, a dramatic change in the Mn magnetic moment was observed through a Mn 3 s multiplet splitting, and via concomitant O 1 s binding energy shifts and other core and valence spectral changes, it was deduced that a high-temperature high-spin phase associated with Jahn–Teller distortion and polaron formation was present. Most of the data obtained in this study could have been obtained with hard X-rays, thus reducing the questions of surface vs. bulk effects that complicated data analysis. These results make it clear that hard X-ray photoemission of core and valence levels will make an important contribution to future studies in nanomagnetism. Extending beyond this capability, atomic structural information, again elementspecific, may be obtained from photoelectron diffraction and holography. Furthermore, spin-polarized photoelectron diffraction and holography would permit determining local magnetic order in complex materials.

Standing Wave Studies Standing wave Studies of Buried Interfaces in the 0.5–10 keV range is Another Desirable Capability

Although many powerful tools for studying solid–vacuum interfaces currently exist (e.g. LEED, STM, AFM, photoemission, surface X-ray diffraction), none of these tools permits a direct probe of the buried solid–solid interfaces that are ubiquitous in nanomaterials. One very promising method for selectively probing such buried interfaces is to use an X-ray standing wave for exciting photoelectrons or fluorescent X-rays, with the excitation strength modulated by a roughly sin2 z variation that may penetrate through the interfac. Strong standing waves are thus created either by diffraction from suitable crystal planes (e.g. in the layered manganites) or from an artificial multilayer structure. Detecting both photoelectrons and fluorescent X-rays would provide greater/ lesser degrees of bulk sensitivity, as well as complementary information on electronic and magnetic structure. Furthermore, a combination of standing-wave excitation techniques with lateral spectromicroscopy would provide truly 3D microscopy.

HOLOGRAPHIC OPPORTUNITIES

X-ray fluorescence holography—sub-Å imaging with no phase problem represents another fruitful frontier. This experiment provides elementspecific local 3D atomic structure in two ways: (i) with a fixed exciting-beam direction by measuring the angular distribution of outgoing fluorescent X-rays as the hologram; or (ii) by carrying out the same experiment in a time-reversed mode, in which the direction of the exciting beam is varied with respect to crystal axes, and the intensity of a given atom's fluorescent X-rays is monitored to yield the hologram. A simple Fourier-transform type of inversion algorithm, most powerful in a multi-energy mode then yields the atomic images.

Various demonstration experiments of this type have been performed, and future possibilities for enhancing image accuracy and dimensions (including magnetic structure) have recently been reviewed. Beyond prior studies, which have demonstrated the element-specific structural capability at the _0.1A Ï level, and have imaged about 40 near-neighbour atoms around a given centre atom, a recent theoretical study suggests that, by going on- and off-resonant scattering conditions for a given element, and then analyzing suitable differences of non-resonant and resonant holograms, the elemental identity of the near-neighbour atoms may be determined, with one experimental study recently seeming to verify this prediction. No other structural probe permits the uniquely identification of the neighbors of a given type of atom.

ADVANCED DETECTOR DEVELOPMENT

Supplemental technological activities that will enable the application of these advanced synchrotron techniques include the development of next-generation, pixilated, ICbased detection systems for both X-rays and electrons. For example, a 1D detector for electrons that has been developed at the Advanced Light Source in Berkeley, could increase count rates by factors of 102–103, permitting time-resolved pump–probe studies in the _200 ms timescale. Beyond this, a large-surface, pixilated, energyresolving X-ray detector would permit time-resolved EXAFS, as well as make X-ray holography a more accessible and more widely utilizable technique. Finally, research and development is needed for a more efficient next-generation approach for electron spin detection, for example, via exchange scattering or spin filtering through a ferromagnetic thin film. Nanomagnetism research carried out with synchrotron radiation would be enormously enhanced by such a spin detector, even if it only increased efficiencies by a factor of 10.

This report attempts to accomplish a number of goals. One is to provide a contemporary glimpse of the field of nanomagnetism. The perspective is necessarily that of the authors. The field is diverse enough that a different collection of authors might frame the picture somewhat differently, but hopefully much of the major interest is included here. We have divided the field into three thematic areas based on three interesting types of materials: patterned nanostructures, clusters, and complex oxides. Another goal was to look into the future and identify the grand challenges that lie ahead.

Of course, such lists of challenges evolve with time as breakthroughs in our understanding emerge. Often the most interesting questions get addressed by parsing them into a different set of questions. As scientists we often answer questions by posing new questions. Having identified where we have been as a field, and where we think we are heading, enables us to plan for the research and instrumentation requirements needed to address the challenges and to probe the difficult questions in the field. This takes the report full circle, because the workshop that served to launch the present report focused on the future needs for hard X-ray synchrotron light sources.

While it is hoped that the present report has value beyond this limited goal, such planning is important in its own right. This is because synchrotron light sources, and major facilities for materials research in general, occupy a significant part of the materials-research funding portfolio. As such, the report makes a number of recommendations to extend the temperature, pressure, and magnetic field characteristics that are accessible at synchrotron beamlines. It also specifically addresses several new techniques and instrumentation directions that are crucial to the next generation of experiments. There is also a desire to have simultaneous access to enhanced optical and electrical support systems within the sample environment. These generic requirements will enhance a range of experiments that exploit the spatial, temporal, and materials realms that can be explored.

Chapter 4

Carbon Nanotube

Carbon nanotubes (CNTs) are allotropes of carbon with a nanostructure that can have a length-to-diameter ratio as large as 28,000,000:1, which is unequalled by any other material. These cylindrical carbon molecules have novel properties that make them potentially useful in many applications in nanotechnology, electronics, optics and other fields of materials science, as well as potential uses in architectural fields. They exhibit extraordinary strength and unique electrical properties, and are efficient conductors of heat. Their final usage, however, may be limited by their potential toxicity.

Nanotubes are members of the fullerene structural family, which also includes the spherical buckyballs. The ends of a nanotube might be capped with a hemisphere of the buckyball structure. Their name is derived from their size, since the diameter of a nanotube is in the order of a few nanometers (approximately 1/50,000th of the width of a human hair), while they can be up to several millimeters in length (as of 2008). Nanotubes are categorized as single-walled nanotubes (SWNTs) and multi-walled nanotubes (MWNTs).

The nature of the bonding of a nanotube is described by applied quantum chemistry, specifically, orbital hybridization. The chemical bonding of nanotubes is composed entirely of sp^2 bonds, similar to those of graphite. This bonding structure, which is stronger than the sp^3 bonds found in diamonds, provides the molecules with their unique strength. Nanotubes naturally align themselves into "ropes" held together by Van der Waals forces. Under high pressure, nanotubes can merge together, trading some sp^2 bonds for sp^3 bonds, giving the possibility of producing strong, unlimited-length wires through high-pressure nanotube linking.

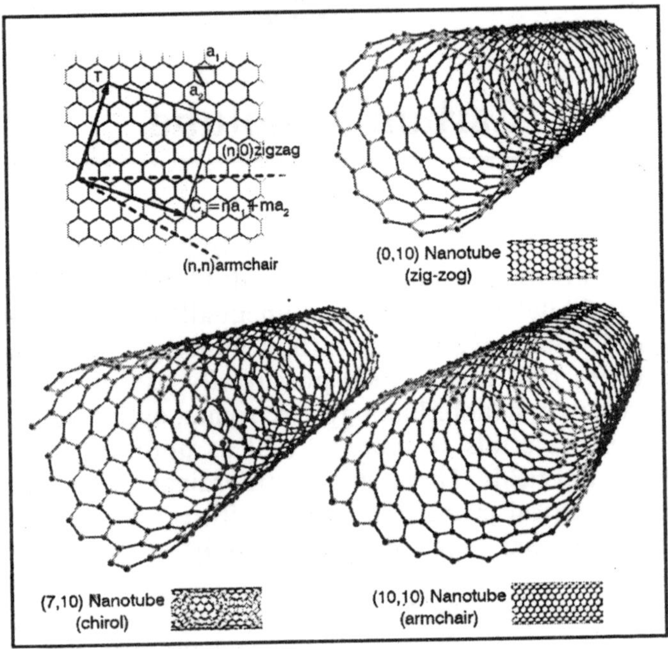

Fig. 3D Model of Three Types of Single-walled Ccarbon Nanotubes.

TYPES OF CARBON NANOTUBES

STRUCTURES OF CARBON NANOTUBES

Single-Walled

Most single-walled nanotubes (SWNT) have a diameter of close to 1nanometer, with a tube length that can be many thousands of times longer. The structure of a SWNT can be conceptualized by wrapping a one-atom-thick layer of graphite called graphene into a seamless cylinder. The way the graphene sheet is wrapped is represented by a pair of indices (*n*,*m*) called the chiral vector. The integers *n* and *m* denote the number of unit vectors along two directions in the honeycomb crystal lattice of graphene. If m=0, the nanotubes are called "zigzag". If n=m, the nanotubes are called "armchair". Otherwise, they are called "chiral".

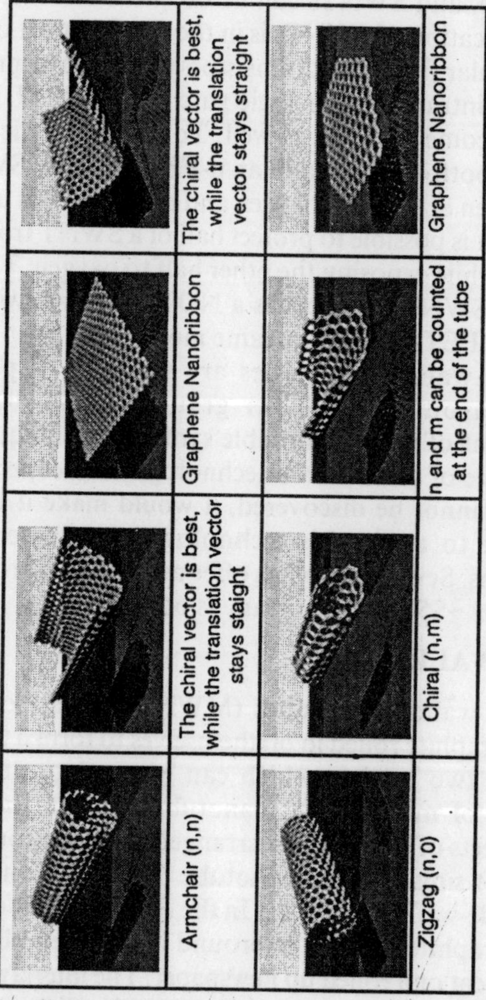

Single-walled nanotubes are a very important variety of carbon nanotube because they exhibit important electric properties that are not shared by the multi-walled carbon nanotube (MWNT) variants. Single-walled nanotubes are the most likely candidate for miniaturizing electronics beyond the micro electromechanical scale that is currently the basis of modern electronics.

The most basic building block of these systems is the electric wire, and SWNTs can be excellent conductors. One useful application of SWNTs is in the development of the first intramolecular field effect transistors (FETs). The production of the first intramolecular logic gate using SWNT FETs has recently become possible as well. To create a logic gate you must have both a p-FET and an n-FET. Because SWNTs are p-FETs when exposed to oxygen and n-FETs when unexposed to oxygen, it is possible to protect half of a SWNT from oxygen exposure, while exposing the other half to oxygen. This results in a single SWNT that acts as a NOT logic gate with both p and n-type FETs within the same molecule.

Single-walled nanotubes are still very expensive to produce, around $1500 per gram as of 2000, and the development of more affordable synthesis techniques is vital to the future of carbon nanotechnology. If cheaper means of synthesis cannot be discovered, it would make it financially impossible to apply this technology to commercial-scale applications. Several suppliers offer as-produced arc discharge SWNTs for ~$50–100 per gram as of 2007.

MULTI-WALLED

Multi-walled nanotubes (MWNT) consist of multiple layers of graphite rolled in on themselves to form a tube shape. There are two models which can be used to describe the structures of multi-walled nanotubes. In the *Russian Doll* model, sheets of graphite are arranged in concentric cylinders, e.g. a (0,8) single-walled nanotube (SWNT) within a larger (0,10) single-walled nanotube. In the *Parchment* model, a single sheet of graphite is rolled in around itself, resembling a scroll of parchment or a rolled up newspaper. The interlayer distance in multi-walled nanotubes is close to the distance between graphene layers in graphite, approximately 3.3 E (330 pm).

The special place of double-walled carbon nanotubes (DWNT) must be emphasized here because they combine very similar morphology and properties as compared to SWNT, while improving significantly their resistance to chemicals. This is especially important when functionalization is required (this means grafting of chemical functions at the surface of the nanotubes) to add new properties to the CNT. In the case of SWNT, covalent functionalization will break some C=C double bonds, leaving "holes" in the structure on the nanotube and thus modifying both its mechanical and electrical properties. In the case of DWNT, only the outer wall is modified. DWNT synthesis on the gram-scale was first proposed in 2003 by the CCVD technique, from the selective reduction of oxides solid solutions in methane and hydrogen.

Fullerite

Fullerites are the solid-state manifestation of fullerenes and related compounds and materials. Being highly incompressible nanotube forms, polymerized single-walled nanotubes (P-SWNT) are a class of fullerites and are comparable to diamond in terms of hardness. However, due to the way that nanotubes intertwine, P-SWNTs don't have the corresponding crystal lattice that makes it possible to cut diamonds neatly. This same structure results in a less brittle material, as any impact that the structure sustains is spread out throughout the material.

Torus

A nanotorus is a theoretically described carbon nanotube bent into a torus (doughnut shape). Nanotori have many unique properties, such as magnetic moments 1000 times larger than previously expected for certain specific radii. Properties such as magnetic moment, thermal stability, etc. vary widely depending on radius of the torus and radius of the tube.

Nanobud

Carbon nanobuds are a newly created material combining two previously discovered allotropes of carbon: carbon nanotubes and fullerenes. In this new material fullerene-like "buds" are covalently bonded to the outer sidewalls of the underlying carbon nanotube. This hybrid material has useful properties of both fullerenes and carbon nanotubes. In particular, they have been found to be exceptionally good field emitters.

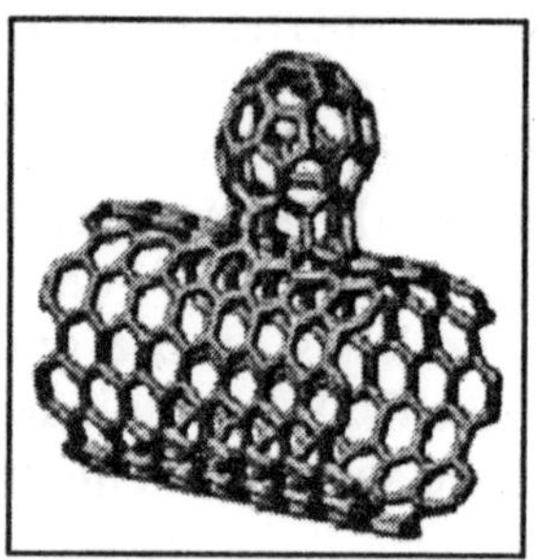

Fig. A Stable Nanobud Structure

In composite materials, the attached fullerene molecules may function as molecular anchors preventing slipping of the nanotubes, thus improving the composite's mechanical properties.

PROPERTIES OF CARBON NANOTUBES

Strength

Carbon nanotubes are the strongest and stiffest materials yet discovered on Earth, in terms of tensile strength and elastic modulus respectively. This strength results from the covalent sp^2 bonds formed between the individual carbon atoms. In 2000, a multi-walled carbon nanotube was tested to have a tensile strength of 63gigapascals (GPa). (This, for illustration, translates into the ability to endure weight of 6300 kg on a cable with cross-section of 1 mm^2.) Since carbon nanotubes have a low density for a solid of 1.3-1.4g•cm"^{3}, its specific strength of up to 48,000kN•m•kg"^{1} is the best of known materials, compared to high-carbon steel's 154kN•m•kg"^{1}.

Under excessive tensile strain, the tubes will undergo plastic deformation, which means the deformation is permanent. This deformation begins at strains of approximately 5% and can increase the maximum strain the tube undergo before fracture by releasing strain energy.

CNTs are not nearly as strong under compression. Because

of their hollow structure and high aspect ratio, they tend to undergo buckling when placed under compressive, torsional or bending stress.

Comparison of Mechanical Properties			
Material	**Young's Modulus (TPa)**	**Tensile Strength (GPa)**	**Elongation at Break (%)**
SWNT	~1 (from 1 to 5)	13-53	16
Armchair SWNT	0.94^T	126.2	23.1
Zigzag SWNT	0.94^T	94.5	15.6-17.5
Chiral SWNT	0.92		
MWNT	0.8-0.9	150	
Stainless Steel	~0.2	~0.65-1	15-50
Kevlar	~0.15	~3.5	~2
Kevlar	0.25	29.6	

The above discussion referred to axial properties of the nanotube, whereas simple geometrical considerations suggest that carbon nanotubes should be much softer in the radial direction than along the tube axis. Indeed, TEM observation of radial elasticity suggested that even the van der Waals forces can deform two adjacent nanotubes. Nanoindentation experiments, performed by several groups on multiwalled carbon nanotubes, , indicated Young's modulus of the order of several GPa confirming that CNTs are indeed rather soft in the radial direction.

Kinetic

Multi-walled nanotubes, multiple concentric nanotubes precisely nested within one another, exhibit a striking telescoping property whereby an inner nanotube core may slide, almost without friction, within its outer nanotube shell thus creating an atomically perfect linear or rotational bearing. This is one of the first true examples of molecular nanotechnology, the precise positioning of atoms to create useful machines. Already this property has been utilized to create the world's smallest rotational motor. Future applications such as a gigahertz mechanical oscillator are also envisaged.

Electrical

Because of the symmetry and unique electronic structure

of graphene, the structure of a nanotube strongly affects its electrical properties. For a given (n, m) nanotube, if $n = m$, the nanotube is metallic; if $n - m$ is a multiple of 3, then the nanotube is semiconducting with a very small band gap, otherwise the nanotube is a moderate semiconductor. Thus all armchair $(n = m)$ nanotubes are metallic, and nanotubes (5, 0), (6, 4), (9, 1), etc. are semiconducting. In theory, metallic nanotubes can carry an electrical current density of 4×10^9 A/cm^2 which is more than 1,000 times greater than metals such as copper.

Thermal

All nanotubes are expected to be very good thermal conductors along the tube, exhibiting a property known as "ballistic conduction," but good insulators laterally to the tube axis. It is predicted that carbon nanotubes will be able to transmit up to 6000 watts per meter per Kelvin at room temperature; compare this to copper, a metal well-known for its good thermal conductivity, which only transmits 385 watts per meter per Kelvin. The temperature stability of carbon nanotubes is estimated to be up to 2800 degrees Celsius in vacuum and about 750 degrees Celsius in air.

Defects

As with any material, the existence of defects affects the material properties. Defects can occur in the form of atomic vacancies. High levels of such defects can lower the tensile strength by up to 85%. Another form of defect that may occur in carbon nanotubes is known as the Stone Wales defect, which creates a pentagon and heptagon pair by rearrangement of the bonds. Because of the very small structure of CNTs, the tensile strength of the tube is dependent on the weakest segment of it in a similar manner to a chain, where a defect in a single link diminishes the strength of the entire chain.

The tube's electrical properties are also affected by the presence of defects. A common result is the lowered conductivity through the defective region of the tube. Some defect formation in armchair-type tubes (which can conduct electricity) can cause the region surrounding that defect to become semiconducting. Furthermore single monoatomic vacancies induce magnetic properties .

The tube's thermal properties are heavily affected by defects. Such defects lead to phonon scattering, which in turn increases the relaxation rate of the phonons. This reduces the mean free path, and reduces the thermal conductivity of nanotube structures. Phonon transport simulations indicate that substitutional defects such as nitrogen or boron will primarily lead to scattering of high frequency optical phonons. However, larger scale defects such as Stone Wales defects cause phonon scattering over a wide range of frequencies, leading to a greater reduction in thermal conductivity.

ONE-DIMENSIONAL TRANSPORT

Due to their nanoscale dimensions, electron transport in carbon nanotubes will take place through quantum effects and will only propagate along the axis of the tube. Because of this special transport property, carbon nanotubes are frequently referred to as "one-dimensional" in scientific articles.

Toxicity

Determining the toxicity of carbon nanotubes has been one of the most pressing questions in Nanotechnology. Unfortunately such research has only just begun and the data are still fragmentary and subject to criticisms. Preliminary results highlight the difficulties in evaluating the toxicity of this heterogeneous material. Parameters such as structure, size distribution, surface area, surface chemistry, surface charge, and agglomeration state as well as purity of the samples, have considerable impact on the reactivity of carbon nanotubes. However, available data clearly show that, under some conditions, nanotubes can cross the membrane barriers and suggests that if raw materials reach the organs they can induce harmful effects as inflammatory and fibrotic reactions.

A study led by Alexandra Porter from the University of Cambridge shows that CNTs can enter human cells and once inside accumulate in the cytoplasm and cause cell death.

Results of rodent studies collectively show that regardless of the process by which CNTs were synthesized and the types and amounts of metals they contained, CNTs were capable of producing inflammation, epithelioid granulomas (microscopic nodules), fibrosis, and biochemical/toxicological changes in the lungs. Comparative toxicity studies in which mice were given equal weights of test materials showed that SWCNTs were more toxic than quartz, which is considered a serious occupational health hazard if it is chronically inhaled. As a control ultrafine carbon black was shown to produce minimal lung responses.

The needle-like fibre shape of CNTs, similar to asbestos fibers, raises fears that widespread use of carbon nanotubes may lead to mesothelioma, cancer of the lining of the lungs caused by exposure to asbestos. A recently published pilot study supports this prediction. Scientists exposed the mesothelial lining of the body cavity of mice, as a surrogate for the mesothelial lining of the chest cavity, to long multiwalled carbon nanotubes and observed asbestos-like, length-dependent, pathogenic behaviour which included inflammation and formation of lesions known as granulomas. Authors of the study conclude that:

"This is of considerable importance, because research and business communities continue to invest heavily in carbon nanotubes for a wide range of products under the assumption that they are no more hazardous than graphite. Our results suggest the need for further research and great caution before introducing such products into the market if long-term harm is to be avoided."

According to co-author, Dr. Andrew Maynard:

"This study is exactly the kind of strategic, highly focused research needed to ensure the safe and responsible development of nanotechnology. It looks at a specific nanoscale material expected to have widespread commercial applications and asks specific questions about a specific health hazard. Even though scientists have been raising concerns about the safety of long, thin carbon nanotubes for over a decade, none of the research needs in the current U.S. federal nanotechnology environment, health and safety risk research strategy address this question." Video commentary

Although further research is required, results presented today clearly demonstrate that, under certain conditions, especially those involving chronic exposure, carbon nanotubes can pose a serious risk to human health.

Synthesis

Techniques have been developed to produce nanotubes in sizeable quantities, including arc discharge, laser ablation, high pressure carbon monoxide (HiPCO), and chemical vapour deposition (CVD). Most of these processes take place in vacuum or with process gases. CVD growth of CNTs can take place in vacuum or at atmospheric pressure. Large quantities of nanotubes can be synthesized by these methods; advances in catalysis and continuous growth processes are making CNTs more commercially viable.

Arc Discharge

Nanotubes were observed in 1991 in the carbon soot of graphite electrodes during an arc discharge, by using a current of 100 amps, that was intended to produce fullerenes. However the first macroscopic production of carbon nanotubes was made in 1992 by two researchers at NEC's Fundamental Research Laboratory. The method used was the same as in 1991. During this process, the carbon contained in the negative electrode sublimates because of the high temperatures caused by the discharge. Because nanotubes were initially discovered using this technique, it has been the most widely used method of nanotube synthesis.

The yield for this method is up to 30 percent by weight and it produces both single- and multi-walled nanotubes with lengths of up to 50 micrometers.

LASER ABLATION

In the laser ablation process, a pulsed laser vaporizes a graphite target in a high temperature reactor while an inert gas is bled into the chamber. The nanotubes develop on the cooler surfaces of the reactor, as the vaporized carbon condenses. A water-cooled surface may be included in the system to collect the nanotubes.

It was invented by Richard Smalley and co-workers at Rice University, who at the time of the discovery of carbon nanotubes, were blasting metals with the laser to produce various metal molecules. When they heard of the discovery[*clarification needed*] they substituted the metals with graphite to create multi-walled carbon nanotubes. Later that year the team used a composite of graphite and metal catalyst particles (the best yield was from a cobalt and nickel mixture) to synthesize single-walled carbon nanotubes.

This method has a yield of around 70% and produces primarily single-walled carbon nanotubes with a controllable diameter determined by the reaction temperature. However, it is more expensive than either arc discharge or chemical vapour deposition.

Chemical Vapour Deposition (CVD)

The catalytic vapour phase deposition of carbon was first reported in 1959, but it was not until 1993 that carbon nanotubes could be formed by this process. In 2007, researchers at the University of Cincinnati (UC) developed a process to grow 18 mm long aligned carbon nanotube arrays on a FirstNano ET3000 carbon nanotube growth system.

During CVD, a substrate is prepared with a layer of metal catalyst particles, most commonly nickel, cobalt, iron, or a combination . he metal nanoparticles can also be produced by other ways, including reduction of oxides or oxides solid solutions. The diameters of the nanotubes that are to be grown are related to the size of the metal particles. This can be controlled by patterned (or masked) deposition of the metal, annealing, or by plasma etching of a metal layer. The substrate is heated to approximately 700°C.

To initiate the growth of nanotubes, two gases are bled into the reactor: a process gas (such as ammonia, nitrogen, hydrogen, etc.) and a carbon-containing gas (such as acetylene, ethylene, ethanol, methane, etc.). Nanotubes grow at the sites of the metal catalyst; the carbon-containing gas is broken apart at the surface of the catalyst particle, and the carbon is transported to the edges of the particle, where it forms the nanotubes. This mechanism is still under discussion. The catalyst particles can stay at the tips of the growing nanotube during the growth process, or remain at the nanotube base, depending on the adhesion between the catalyst particle and the substrate.

CVD is a common method for the commercial production of carbon nanotubes. For this purpose, the metal nanoparticles will be carefully mixed with a catalyst support (e.g., MgO, Al2O3, etc) to increase the specific surface area for higher yield of the catalytic reaction of the carbon feedstock with the metal particles. One issue in this synthesis route is the removal of the catalyst support via an acid treatment, which sometimes could destroy the original structure of the carbon nanotubes. However, alternative catalyst supports that are soluble in water have been shown to be effective for nanotube growth.

If a plasma is generated by the application of a strong electric field during the growth process (plasma enhanced chemical vapour deposition*), then the nanotube growth will follow the direction of the electric field. By properly adjusting the geometry of the reactor it is possible to synthesize vertically aligned carbon nanotubes (i.e., perpendicular to the substrate), a morphology that has been of interest to researchers interested in the electron emission from nanotubes. Without the plasma, the resulting nanotubes are often randomly oriented. Under certain reaction conditions, even in the absence of a plasma, closely spaced nanotubes will maintain a vertical growth direction resulting in a dense array of tubes resembling a carpet or forest.

Of the various means for nanotube synthesis, CVD shows the most promise for industrial scale deposition in terms of its price/unit ratio. There are additional advantages to the CVD synthesis of nanotubes. Unlike the above methods, CVD is capable of growing nanotubes directly on a desired substrate, whereas the nanotubes must be collected in the other growth techniques. The growth sites are controllable by careful deposition of the catalyst. In 2007, a team from Meijo University has shown a high-efficiency CVD technique for growing carbon nanotubes from camphor.

A team of researchers at Rice University, until recently led by the late Dr. Richard Smalley, has concentrated upon finding methods to produce large, pure amounts of particular types of nanotubes. Their approach grows long fibers from many small seeds cut from a single nanotube; all of the resulting fibers were found to be of the same diameter as the original nanotube and are expected to be of the same type as the original nanotube. Further characterization of the resulting nanotubes and improvements in yield and length of grown tubes are needed. CVD growth of multi-walled nanotubes is used by several companies to produce materials on the ton scale, including NanoLab, Bayer, Arkema, Nanocyl, Nanothinx, Hyperion Catalysis, Mitsui, and Showa Denko.

Natural, Incidental, and Controlled Flame Environments

Fullerenes and carbon nanotubes are not necessarily products of high-tech laboratories; they are commonly formed in such mundane places as ordinary flames, produced by burning methane, ethylene, and benzene, and they have been found in soot from both indoor and outdoor air.

However, these naturally occurring varieties can be highly irregular in size and quality because the environment in which they are produced is often highly uncontrolled. Thus, although they can be used in some applications, they can lack in the high degree of uniformity necessary to meet many needs of both research and industry. Recent efforts have focused on producing more uniform carbon nanotubes in controlled flame environments. Nano-C, Inc of Westwood, Massachusetts, is producing flame synthesized single-walled carbon nanotubes. This method has promise for large scale, low cost nanotube synthesis, though it must compete with rapidly developing large scale CVD production.

Potential and Current Applications

The strength and flexibility of carbon nanotubes makes

them of potential use in controlling other nanoscale structures, which suggests they will have an important role in nanotechnology engineering. The highest tensile strength an individual multi-walled carbon nanotube has been tested to be is 63GPa.

A 2006 study published in *Nature* determined that some carbon nanotubes are present in Damascus steel, possibly helping to account for the legendary strength of the (almost ancient) swords made of it.

Structural

Because of the great mechanical properties of the carbon nanotubule, a variety of structures have been proposed ranging from everyday items like clothes and sports gear to combat jackets and space elevators. However, the space elevator will require further efforts in refining carbon nanotube technology, as the practical tensile strength of carbon nanotubes can still be greatly improved.

For perspective, outstanding breakthroughs have already been made. Pioneering work led by Ray H. Baughman at the NanoTech Institute has shown that single and multi-walled nanotubes can produce materials with toughness unmatched in the man-made and natural worlds.

Recent research by James D. Iverson and Brad C. Edwards has revealed the possibility of cross-linking CNT molecules prior to incorporation in a polymer matrix to form a super high strength composite material. This CNT composite could have a tensile strength on the order of 20 million psi (138 GPa, for 106 MN•m•kg"1), potentially revolutionizing many aspects of engineering design where low weight and high strength is required.[citation needed]

In Electrical Circuits

Carbon nanotubes have many properties—from their unique dimensions to an unusual current conduction mechanism—that make them ideal components of electrical circuits. For example, they have shown to exhibit strong electron-phonon resonances, which indicate that under certain direct current (dc) bias and doping conditions their current and the average electron velocity, as well as the electron concentration on the tube oscillate at terahertz frequencies. These resonances can be used to make terahertz sources or sensors.

Nanotube based transistors have been made that operate at room temperature and that are capable of digital switching using a single electron.

One major obstacle to realization of nanotubes has been the lack of technology for mass production. However, in 2001 IBM researchers demonstrated how nanotube transistors can be grown in bulk, not very differently from silicon transistors. The process they used is called "constructive destruction" which includes the automatic destruction of defective nanotubes on the wafer.

This has since then been developed further and single-chip wafers with over ten billion correctly aligned nanotube junctions have been created. In addition it has been demonstrated that incorrectly aligned nanotubes can be removed automatically using standard photolithography equipment.

The first nanotube integrated memory circuit was made in 2004. One of the main challenges has been regulating the conductivity of nanotubes. Depending on subtle surface features a nanotube may act as a plain conductor or as a semiconductor. A fully automated method has however been developed to remove non-semiconductor tubes.

Most recently, collaborating American and Chinese researchers at Duke University and Peking University announced a new CVD recipe involving a combination of ethanol and methanol gases and quartz substrates resulting in horizontally aligned arrays of 95-98% semiconducting nanotubes. This is considered a large step towards the ultimate goal of producing perfectly aligned, 100% semiconducting carbon nanotubes for mass production of electronic devices.

An alternative way to make transistors out of carbon nanotubes has been to use random networks of them. By doing so one averages all of their electrical differences and one can produce devices in large scale at the wafer level. This approach was first patented by Nanomix Inc.(date of original application in June 2002). It was first published in the academic literature by the Naval Research Laboratory in 2003 through independent research work. This approach also enabled Nanomix to make the first transistor on a flexible and transparent substrate.,

Nanotubes are usually grown on nanoparticles of magnetic metal (Fe, Co) that facilitates production of electronic (spintronic) devices. In particular control of current through a field-effect transistor by magnetic field has been demonstrated in such a single-tube nanostructure.

As a Vessel for Drug Delivery

The nanotube's versatile structure allows it to be used for a variety of tasks in and around the body. Although often seen especially in cancer related incidents, the carbon nanotube is often used as a vessel for transporting drugs into the body. The nanotube allows for the drug dosage to hopefully be lowered by localizing its distribution, as well as significantly cut costs to pharmaceutical companies and their consumers. The nanotube commonly carries the drug one of two ways: the drug can be attached to the side or trailed behind, or the drug can actually be placed inside the nanotube. Both of these methods are effective for the delivery and distribution of drugs inside of the body.

Current Applications

They are used as bulk nanotubes, which is a mass of rather unorganized fragments of nanotubes. Bulk nanotube materials may never achieve a tensile strength similar to that of individual tubes, but such composites may nevertheless yield strengths sufficient for many applications. Bulk carbon nanotubes have already been used as composite fibers in polymers to improve the mechanical, thermal and electrical properties of the bulk product.

Easton Bicycle Components have been in partnership with Zyvex, using CNT technology in a number of their components

- including flat and riser handlebars, cranks, forks, seatposts, stems and aero bars.

SOLAR CELLS

The solar cell developed at NJIT uses a carbon nanotubes complex, formed by carbon nanotubes and combines them with tiny carbon buckyballs (known as fullerenes) to form snake-like structures. Buckyballs trap electrons, although they can't make electrons flow. Add sunlight to excite the polymers, and the buckyballs will grab the electrons. Nanotubes, behaving like copper wires, will then be able to make the electrons or current flow.

Ultracapacitors

MIT Laboratory for Elecromagnetic and Electronic Systems uses nanotubes to improve ultracapacitors. The activated charcoal used in conventional ultracapacitors has many small hollow spaces with a distribution of sizes, which create together a large surface to store electric charges. But as charge is quantized into elementary charges, i.e. electrons, and each of these needs a minimum space, a large fraction of the electrode surface is not available for storage because the hollow spaces are too small. With an electrode made out of nanotubes, the spaces are hoped to be tailored to size - few too large or too small - and consequently the capacity is hoped to be increased considerably.

Other Applications

Carbon nanotubes have also been implemented in nanoelectromechanical systems, including mechanical memory elements (NRAM being developed by Nantero Inc.) and nanoscale electric motors (see Nanomotor).

Carbon nanotubes have also been proposed as a possible gene delivery vehicle and for use in combination with radiofrequency fields to destroy cancer cells.

In May 2005, Nanomix Inc has put on the market an electronic device - a Hydrogen sensor - that integrated carbon nanotubes on a silicon platform. Since then Nanomix has been patenting many such sensor applications such as in the field of carbon dioxide, nitrous oxide, glucose, DNA detection etc.

Eikos Inc of Franklin, Massachusetts and Unidym Inc. of Silicon Valley, California are developing transparent, electrically conductive films of carbon nanotubes to replace indium tin oxide (ITO). Carbon nanotube films are substantially more mechanically robust than ITO films, making them ideal for high reliability touch screens and flexible displays. Printable water-based inks of carbon nanotubes are desired to enable the production of these films to replace ITO. Nanotube films show promise for use in displays for computers, cell phones, PDAs, and ATMs. A nanoradio, a radio receiver consisting of a single nanotube, was demonstrated in 2007. In 2008 it was shown that a sheet of nanotubes can operate as a loudspeaker if an alternating current is applied. The sound is not produced through vibration but thermoacoustically.

Carbon nanotubes are said to have the strength of diamond, and research is being made into weaving them into clothes to create stab-proof and bulletproof clothing. The nanotubes would effectively stop the bullet from penetrating the body but the kinetic energy of the bullet would be likely to cause broken bones and internal bleeding. A flywheel made of carbon nanotubes could be spun at extremely high velocity on a floating magnetic axis, and potentially store energy at a density approaching that of conventional fossil fuels. Since energy can be added to and removed from flywheels very efficiently in the form of electricity, this might offer a way of storing electricity, making the electrical grid more efficient and variable power suppliers (like wind turbines) more useful in meeting energy needs. The practicality of this depends heavily upon the cost of making massive, unbroken nanotube structures, and their failure rate under stress.

Rheological properties can also be shown very effectively by carbon nanotubes.

REGULATION

In October 2008, the Department of Toxic Substances

Control (DTSC), within the California Environmental Protection Agency, announced its intent to request information regarding analytical test methods, fate and transport in the environment, and other relevant information from manufacturers of carbon nanotubes. The term "manufacturers" includes persons and businesses that produce nanotubes in California, or import carbon nanotubes into California for sale. The purpose of this information request will be to identify information gaps and to develop information about carbon nanotubes, an important emerging nanomaterial.

DTSC is exercising its' authority under California Health and Safety Code, Chapter 699, sections 57018-57020. These sections were added as a result of the adoption of Assembly Bill AB 289 (2006). They are intended to make information on the fate and transport, detection and analysis, and other information on chemicals more available. The law places the responsibility to provide this information to the Department on those who manufacture or import the chemicals.

DISCOVERY

A 2006 editorial written by Marc Monthioux and Vladimir Kuznetsov in the journal *Carbon* has described the interesting and often misstated origin of the carbon nanotube. A large percentage of academic and popular literature attributes the discovery of hollow, nanometer sized tubes composed of graphitic carbon to Sumio Iijima of NEC in 1991.

L. V. Radushkevich and V. M. Lukyanovich published clear images of 50 nanometer diameter tubes made of carbon in the Soviet *Journal of Physical Chemistry*. This discovery was largely unnoticed, as the article was published in the Russian language, and Western scientists' access to Soviet press was limited during the Cold War. It is likely that carbon nanotubes were produced before this date, but the invention of the transmission electron microscope allowed the direct visualization of these structures.

Carbon nanotubes have been produced and observed under a variety of conditions prior to 1991. A paper by Oberlin, Endo, and Koyama published in 1976 clearly showed hollow carbon fibers with nanometer-scale diameters using a vapour-growth technique. Additionally, the authors show a TEM image of a nanotube consisting of a single wall of graphene. Later, Endo has referred to this image as a single-walled nanotube.

Furthermore, in 1979, John Abrahamson presented evidence of carbon nanotubes at the 14th Biennial Conference of Carbon at Penn State University. The conference paper described carbon nanotubes as carbon fibers which were produced on carbon anodes during arc discharge. A characterization of these fibers was given as well as hypotheses for their growth in a nitrogen atmosphere at low pressures.

In 1981 a group of Soviet scientists published the results of chemical and structural characterization of carbon nanoparticles produced by a thermocatalytical disproportionation of carbon monoxide. Using TEM images and XRD patterns, the authors suggested that their "carbon multi-layer tubular crystals" were formed by rolling graphene layers into cylinders. Additionally, they speculated that during rolling graphene layers into a cylinder, many different arrangements of graphene hexagonal nets are possible. They suggested two possibilities of such arrangements: circular arrangement (armchair nanotube) and a spiral, helical arrangement (chiral tube).

In 1987, Howard G. Tennent of Hyperion Catalysis was issued a U.S. patent for the production of "cylindrical discrete carbon fibrils" with a "constant diameter between about 3.5 and about 70nanometers..., length 10^2 times the diameter, and an outer region of multiple essentially continuous layers of ordered carbon atoms and a distinct inner core...."

Iijima's discovery of multi-walled carbon nanotubes in the insoluble material of arc-burned graphite rods and Mintmire, Dunlap, and White's independent prediction that if single-walled carbon nanotubes could be made, then they would exhibit remarkable conducting properties helped create the initial buzz that is now associated with carbon nanotubes. Nanotube research accelerated greatly following the independent discoveries by Bethune at IBM and Iijima at NEC of *single-walled* carbon nanotubes and methods to specifically produce them by adding transition-metal catalysts to the carbon in an arc discharge.

The arc discharge technique was well-known to produce the famed Buckminster fullerene on a preparative scale, and these results appeared to extend the run of accidental discoveries relating to fullerenes. The original observation of fullerenes in mass spectrometry was not anticipated, and the first mass-production technique by Krδtschmer and Huffman was used for several years before realizing that it produced fullerenes.

The discovery of nanotubes remains a contentious issue, especially because several scientists involved in the research could be likely candidates for the Nobel Prize. Many believe that Iijima's report in 1991 is of particular importance because it brought carbon nanotubes into the awareness of the scientific community as a whole. See the reference for a review of the history of the discovery of carbon nanotubes.

Similar to the matter of nanotube discovery, the question what is the thinnest carbon nanotube is a matter of debate. The possible candidates can be given as follows: Nanotubes of diameter about 0.40 nm have been reported in 2000 literally on the same page of the journal Nature; however, they are not free standing, but enclosed in zeolite crystals or are innermost shells of the multi-wall nanotubes. Later, inner shells of MWNTs of only 0.3 nm in diameter have been reported. The thinnest free-standing nanotube, by September 2003, has diameter of 0.43 nm.

Chapter 5

Conducting Polymers

Until about 30 years ago all carbon based polymers were rigidly regarded as insulators. The idea that plastics could be made to conduct electricity would have been considered to be absurd. Indeed, plastics have been extensively used by the electronics industry because of this very property. They were utilized as inactive packaging and insulating material. This very narrow perspective is rapidly changing as a new class of polymer known as intrinsically conductive polymer or electroactive polymers are being discovered.

Although this class is in its infancy, much like the plastic industry was in the 30's and 50's, the potential uses of these are quite significant. In 1958, polyacetylene was first synthesised by Natta et al. as a black powder. This was found to be a semi-conductor with a conductivity between 7×10^{-11} to 7×10^{-3} Sm^{-1}, depending upon how the polymer was processed and manipulated. This compound remained a scientific curiosity until 1967, when a postgraduate student of Hideki Shirakawa at the Tokyo Institute of Technology was attempting to synthesise polyacetylene, and a silvery thin film was produced as a result of a mistake.

It was found that 1000 times too much of the Ziegler-Natta catalyst, $Ti(O\text{-}n\text{-}But)^4$ - Et_3Al, had been used. When this film was investigated it was found to be semiconducting, with a similar level of conductivity to the best of the conducting black powders. Further investigations, initially aimed to produce thin films of graphite, showed that exposure of this form of polyacetylene to halogens increased its conductivity a billion fold. Undoped, the polymer was silvery, insoluble and intractable, with a conductivity similar to that of semiconductors. When it was weakly oxidised by compounds such as iodine it turned a golden colour and its conductivity increased to about 10^4 Sm^{-1}.In the 1980's polyheterocycles were first developed.

Polyheterocycles were found to be much more air stable than polyacetylene, although their conductivities were not so high, typically about 10^3 Sm^{-1}. By adding various side groups to the polymer backbone, derivatives which were soluble in various solvents were prepared. Other side groups affected properties such as their colour and their reactivity to oxidising and reducing agents. A Logarithmic conductivity ladder of some of these polymers are shown below.

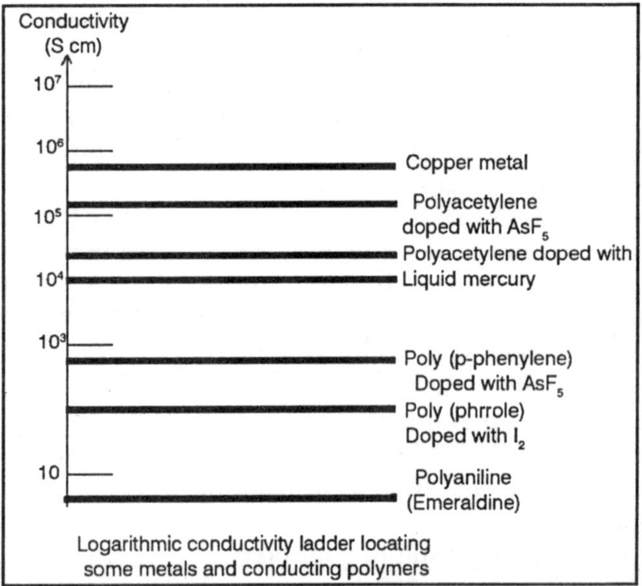

Logarithmic conductivity ladder locating some metals and conducting polymers

Since then it has been found that about a dozen different polymers and polymer derivatives undergo this transition when doped with a weak oxidation agent or reducing agent. They are all various conjugated polymers. This early work has led to an understanding of the mechanisms of charge storage and charge transfer in these system. All have a highly conjugated electronic state.

This also causes the main problems with the use of these systems, that of processibility and stability. Most early conjugated polymers were unstable in air and were not capable of being processed. The most recent research in this has been the development of highly conducting polymers with good stability and acceptable processing attributes.

CHARGE STORAGE

One early explanation of conducting polymers used band theory as a method of conduction. This said that a half filled valence band would be formed from a continuous delocalized p -system. This would be an ideal condition for conduction of electricity. However, it turns out that the polymer can more efficiently lower its energy by bond alteration (alternating short and long bonds), which, introduces a band width of 1.5 ev making it a high energy gap semiconductor.

The polymer is transformed into a conductor by doping it with either an electron donator or an electron acceptor. This is reminiscent of doping of silicon based semiconductors where silicon is doped with either arsenic or boron. However, while the doping of silicon produces a donor energy level close to the conduction band or a acceptor level close to the valence band, this is not the case with conducting polymers. The evidence for this is that the resulting polymers do not have a high enough concentration of free spins, as determined by electron spin spectroscopy. Initially the free spins concentration increases with concentration of dopant.

At larger concentrations, however, the concentration of free spins levels of at a maximum. To understand this it is necessary to examine the way in which charge is stored along the polymer chain and its effect. The polymer may store charge in two ways. In an oxidation process it could either lose an electron from one of the bands or it could localize the charge over a small section of the chain. Localizing the charge causes a local distortion due a change in geometry, which costs the polymer some energy. However, the generation of this local geometry decreases the ionization energy of the polymer chain and increases its electron affinity making it more able to accommodate the newly formed charges. This method increases the energy of the polymer less than it would if the charge was delocalized and, hence, takes place in preference of charge delocalization. This is consistent with an increase in disorder detected after doping by Raman spectroscopy. A similar scenario occurs for a reductive process. Typical oxidizing dopants used include iodine, arsenic pentachloride, iron(III) chloride and $NOPF_6$. A typical reductive dopant is sodium naphthalide. The main criteria is its ability to oxidize or reduce the polymer without lowering its stability or whether or not they are capable of initiating side reactions that inhibit the polymers ability to conduct electricity.

An example of the latter is the doping of a conjugated polymer with bromine. Bromine it too powerful an oxidant and adds across the double bonds to form sp3 carbons. The same problem may also occur with $NOPF_6$ if left too long. The oxidative doping of polypyrrole proceeds in the following way. An electron is removed from the psystem of the backbone producing free radical and a spinless positive charge. The radical and cation are coupled to each other via local resonance of the charge and the radical. In this case, a sequence of quinoid-like rings is used.

The distortion produced by this is of higher energy than the remaining portion of the chain. The creation and separation of these defects costs a considerable amount of energy. This limits the number of quinoid-like rings that can link these two bound species together. In the case of polypyrrole it is believed that the lattice distortion extends over four pyrrole rings. This combination of a charge site and a radical is called a polaron. This could be either a radical cation or radical anion. This creates a new localized electronic states in the gap, with the lower energy states being occupied by a single unpaired electrons. The polaron state of polypryyole are symmetrically located about 0.5 eV from the band edges.

Upon further oxidation the free radical of the polaron is removed, creating a new spinless defect called a bipolaron. This is of lower energy than the creation of two distinct polarons. At higher doping levels it become possible that two polarons combine to form a bipolaron. Thus at higher doping levels the polarons are replaced with bipolarons. The bipolarons are located symmetrically with a band gap of 0.75 eV for polypryyole. This eventually, with continued doping, forms into a continuous bipolaron bands. Their band gap also increases as newly formed bipolarons are made at the expense of the band edges. For a very heavily dope polymer it is conceivable that the upper and the lower bipolaron bands will merge with the conduction and the valence bands respectively to produce partially filled bands and metallic like conductivity. This is shown below.

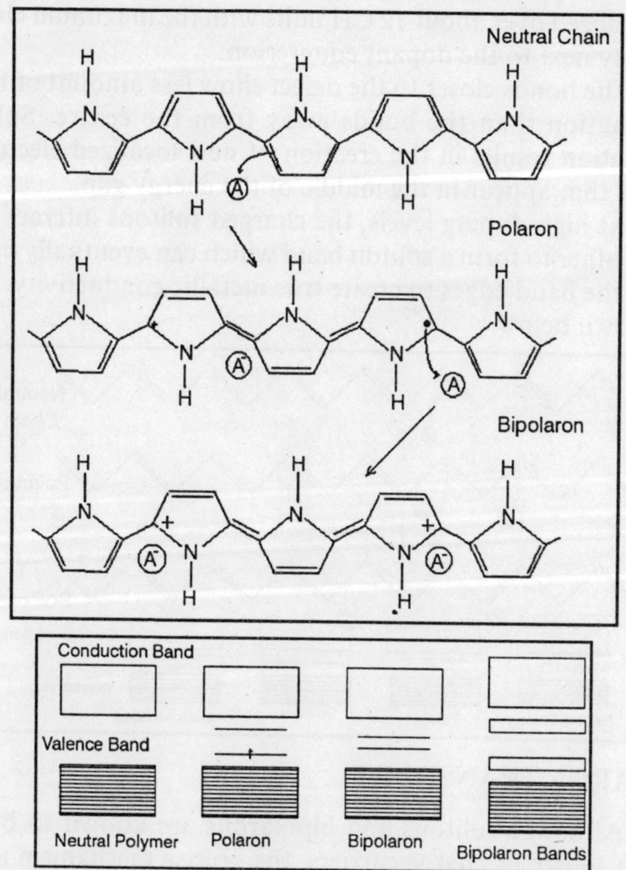

Conjugated polymers with a degenerate ground state have

a slightly different mechanism. As with polypryyole, polarons and bipolarons are produced upon oxidation. However, because the ground state structure of such polymers are twofold degenerate, the charged cation are not bound to each other by a higher energy bonding configuration and can freely separate along the chain.

The effect of this is that the charged defects are independent of one another and can form domain walls that separate two phases of opposite orientation and identical energy. These are called solitons and can sometimes be neutral. Solitons produced in polyacetylene are believed to be delocalized over about 12 CH units with the maximum charge density next to the dopant counterion.

The bonds closer to the defect show less amount of bond alternation than the bonds away from the centre. Soliton formation results in the creation of new localized electronic states that appear in the middle of the energy gap.

At high doping levels, the charged solitons interact with each other to form a soliton band which can eventually merge with the band edges to create true metallic conductivity. This is shown below.

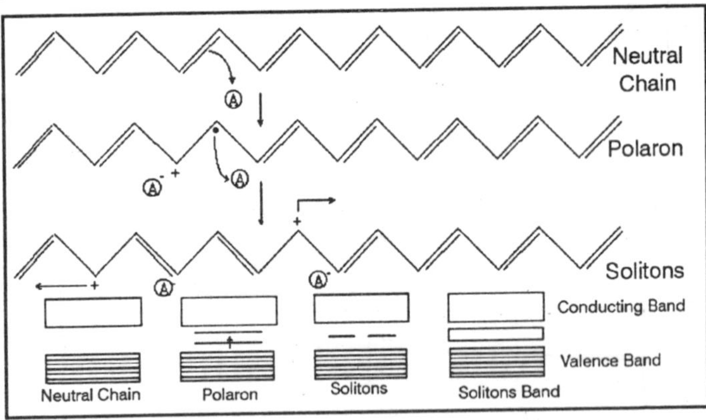

CHARGE TRANSPORT

Although solitons and bipolarons are known to be the main source of charge carriers, the precise mechanism is not yet fully understood. The problem lies in attempting to trace the path of the charge carriers through the polymer. All of these polymers are highly disordered, containing a mixture of crystalline and amorphous regions. It is necessary to consider the transport along and between the polymer chains and also the complex boundaries established by the multiple number of phases.

This has been studied by examining the effect of doping, of temperature, of magnetism and the frequency of the current used. These test show that a variety of conduction mechanisms are used. The main mechanism used is by movement of charge carriers between localized sites or between soliton, polaron or bipolaron states. Alternatively, where inhomogeneous doping produces metallic island dispersed in an insulating matrix, conduction is by movement of charge carriers between highly conducting domains. Charge transfer between these conducting domains also occurs by thermally activated hopping or tunnelling. This is consistent with conductivity being proportional to temperature.

STABILITY

There are two distinct types of stability. Extrinsic stability is related to vulnerability to external environmental agent such as oxygen, water, peroxides. This is determined by the polymers susceptibility of charged sites to attack by nucleophiles, electrophiles and free radicals. If a conducting polymer is Extrinsic unstable then it must be protected by a stable coating. Many conducting polymers, however, degrade over time even in dry, oxygen free environment.

This intrinsic instability is thermodynamic in origin. It is likely to be cause by irreversible chemical reaction between charged sites of polymer and either the dopant counter ion or the p-system of an adjacent neutral chain, which produces an sp^3 carbon, breaking the conjugation. Intrinsic instability can also come from a thermally driven mechanism which causes the polymer to lose its dopant. This happens when the charge sites become unstable due to conformational changes in the polymer backbone. This has been observed in alkyl substituted polythiophenes.

PROCESSIBILITY

Conjugated polymers may be made by a variety of techniques, including cationic, anionic, radical chain growth, co-ordination polymerization, step growth polymerization or electrochemical polymerization. Electrochemical polymerization occurs by suitable monomers which are electrochemically oxidized to create an active monomeric and dimeric species which react to form a conjugated polymer backbone. The main problem with electrically conductive plastics stems from the very property that gives it its conductivity, namely the conjugated backbone.

This causes many such polymers to be intractable, insoluble films or powders that cannot melt. There are two main strategies to overcoming these problems. There are to either modify the polymer so that it may be more easily processed, or to manufacture the polymer in its desired shape and form. There are, at this time, four main methods used to achieve these aims. The first method is to manufacture a malleable polymer that can be easily converted into a conjugated polymer. This is done when the initial polymer is in the desired form and then, after conversion, is treated so that it becomes a conductor.

The treatment used is most often thermal treatment. The precursor polymer used is often made to produce highly aligned polymer chain which are retained upon conversion. These are used for highly orientated thin films and fibres. Such films and fibres are highly anisotropic, with maximum conductivity along the stretch direction. The second method is the synthesis of copolymers or derivatives of a parent conjugated polymer with more desirable properties. This method is the more traditional one for making improvements to a polymer.

What is done is to try to modify the structure of the polymer to increase its processibility without compromising its conductivity or its optical properties. All attempts to do this on polyacetylene have failed as they always significantly reduced its conductivity. However, such attempts on polythiophenes and polypyrroles proved more fruitful. The hydrogen on carbon 3 on the thiophene or the pyrrole ring was replaced with an alkyl group with at least four carbon atoms in it.

The resulting polymer, when doped, has a comparable conductivity to its parent polymer whilst be able to melt and it is soluble. A water soluble version of these polymers has been produced by placing carboxylic acid group or sulphonic acid group on the alkyl chains. If sulphonic acid group groups are used along with built-in ionizable groups. then such system can maintain charge neutrality in its oxidized state and so they effectively dope themselves. Such polymers are referred to as "self-doped" polymers. One of the most highly conductive derivative of polythiophene is made by replacing the hydrogen on carbon three with a $-CH_2-O-CH_2CH_2-O-CH_2CH_2-O-CH_3$.

This is soluble and reaches a conductivity of about 1000 S cm^{-1} upon doping. The third method is to grow the polymer into its desired shape and form. An insulating polymer impregnated with a catalyst system is fabricated into its desired form. This is then exposed to the monomer, usually a gas or a vapour. The monomer then polymerizes on the surface of the insulating plastic producing a thin film or a fibre. This is then doped in the usual manner.

A variation of this technique is electrochemical polymerization with the conducting polymer being deposited on an electrode either the polymerization stage or before the electrochemical polymerization. This cast may be used for further processing of the conducting polymer. For instance, by stretching aligned bends of polyacetylene/polybutadiene the conductivity increase 10 fold, due to the higher state of order produced by this deformation. The final method is the use of Langmuir-Blodgett trough to manipulate the surface active molecules into a highly ordered thin films whose structure and thickness which are controllable at the molecular layer.

Amphiphilic molecules with hydrophilic and hydrophobic groups produces monolayers at the air-water surface interface of a Langmuir-Blodgett trough. This is then transferred to a substrate creating a multilayer structure comprised of molecular stacks which are normal about 2.5 nm thick. This is a development from the creation of insulating films by the same technique.

The main advantage of this technique is its unique ability to allow control over the molecular architecture of the conducting films produced.

It can be used to create complex multilayer structures of functionally different molecular layers as determined by the chemist.

By producing alternating layers of conductor and insulator it is possible to produce highly anisotropic film which is conducting within the plane of the film, but insulating across it.

Table. Stability And Processing Attributes of Some Conducting Polymers

Polymer	Conductivity (Ω^{-1}cm)	Stability (doped state)	Processing Possibilities
Polyacetylene	10^3-10^5	Poor	Limited
Polyphenylene	1000	Poor	Limited
PPS	100	Poor	Excellent
PPV	1000	Poor	Limited
Polypyrroles	100	Good	Good
Polythiophenes	100	Good	Excellent
Polyaniline	10	Good	Good

APPLICATIONS

The extended p-systems of conjugated polymer are highly susceptible to chemical or electrochemical oxidation or reduction.

These alter the electrical and optical properties of the polymer, and by controlling this oxidation and reduction, it is possible to precisely control these properties. Since these reactions are often reversible, it is possible to systematically control the electrical and optical properties with a great deal of precision. It is even possible to switch from a conducting state to an insulating state. There are two main groups of applications for these polymers. The first group utilizes their conductivity as its main property. The second group utilizes this electroactivity. They are shown below:

Group 1	Group2
Electrostatic materials	Molecular Electronics
Conducting Adhesives	Electrical Displays
Electromagnetic shielding	Chemical Biochemical and thermal sensors
Printed circuit boards	Rechargeable batteries and solid electrolytes
Artifucial clothing	Optical computers
Piezoceramics	Ion exchange membranes
Active electronics (diodes transistors)	Electromechanical actuators
Aircraft structures	'Smart' structures
	Switches

Much research will be needed before many of the above applications will become a reality. The stability and processibility both need to be substantially improved if they are to be used in the market place. The cost of such polymers must also be substantially lowered. However, one must consider that, although conventional polymers were synthesized and studied in laboratories around the world, they did not become widespread until years of research and development had been done.

In a way, conducting polymers are at the same stage of development as their insulating brothers were some 50 years ago. Regardless of the practical application that are eventually developed for them, they will certainly challenge researchers in the years to come with new and unexpected phenomena. Only time will tell whether the impact of these novel plastics will be as large as their insulating relatives.3

Chapter 6

Digital Recording Techniques

In the simplest case, the word digital refers to the representation of a quantity in numerical form and analog refers to a continuous physical quantity. To digitize means to convert an analog physical quantity into a numerical value. For example, if we represent the intensity of a sound by numbers proportionally related to the intensity, the analog value of the intensity has been represented digitally.

The accuracy of the digital conversion depends upon the number of discrete numerical values that can be assigned and the rate at which these numerical measurements are made.

For example, 4 numerical levels will represent changes in the amplitude of sound less accurately than 256 numerical levels and a rate of 8 conversion/sec will be less accurate than a rate of 10,000 conversions/sec.

The process for digitally coding sound by computer was first developed in 1957 by Max Mathews of Bell Telephone Laboratories in Murray Hill (Mathews, 1963). Other advances in digital electronics and microchips led to the development of the first digital Pulse Code Modulation (PCM) audio recorder in 1967 at the NHK Technical Research Institute (Nakajima, 1983).

This machine was a 12- bit companded scheme (using a compression/expansion of sound to improve dynamic range) with a 30 kHz sampling rate.

Data were recorded on a one- track, two-head helical scan VTR (Video Tape Recorder). The first commercial PCM/ digital recording session was performed by DENON in 1972 (Takeaki, 1989).

DIGITAL RECORDING PRINCIPLES

During digital recording of the analog signal, analog to digital (A/D) conversion takes place from continuous time-amplitude coordinates to discrete time-amplitude coordinates as illustrated in figure. The difference between the instantaneous analog signal and de digital representation is digital error.

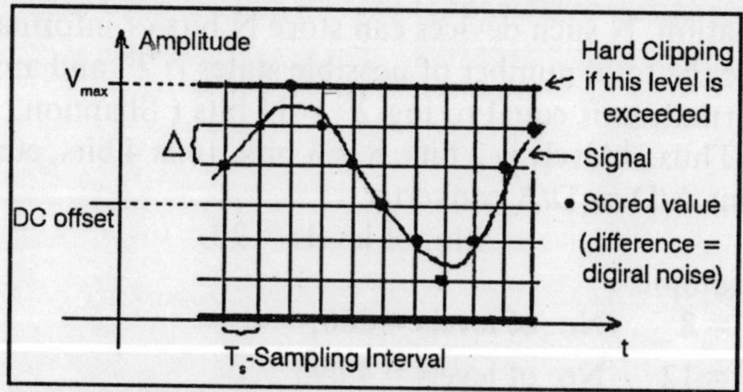

Fig. Use of an A/D (or D/A) converter to convert a continuous function (time-amplitude) to a discrete function (discrete time - discrete amplitude). Conversion introduces a digital error in the signal - digital noise.

We will separately consider the consequences of discrete time and discrete amplitude coordinates on the representation of the analog signal.

Discrete Time

Nyquist theorem. The Nyquist theorem states that if a signal V(t) does not contain frequencies higher than $f_s/2$ (where $f_s = 1/T_s$), then it can be fully recovered from its sampled values $V(nT_s)$ at discrete times $t_n = nT_s$ where n =... -1, 0, 1, 2, 3...

$$V(t) = \sum_{n=\infty}^{\infty} V(n \cdot T_s) \frac{\sin\left[\pi \cdot f_s (t - n \cdot T_s)\right]}{\pi \cdot f_s \cdot (t - n \cdot T_s)}$$

where:

$f_s = 1/T_s$, the sampling frequency

V(t) = value of signal at arbitrary time t.

This is a remarkable result. The recovered signal will have all the frequencies in the range from 0 to $f_s/2$ Hz.

Discrete Amplitude

The term bit stands for binary digit and is associated with a two-choice situation (0 and 1). Thus, any digital system with just two levels has a 1 bit resolution. Generally, the logarithm to the base 2 is used to convert the number of available quantization levels to number of bits. A device with two stable positions, such as a relay or a flip-flop, can store 1 bit of information. N such devices can store N bits of information, because the total number of possible states is 2^N and amount of information is equal to $\log_2 2^N = N$ bits (Shannon, 1949/ 1975). Thus, 4 levels is 2 bits, 8 is 3 bits, 16 is 4 bits, etc. For an N-bit A/D or D/A converter

$$\text{No. of levels} = 2^N$$

Example:

N = 8	No. of levels = 256
N = 12	No. of levels = 4,096
N = 16	No. of levels = 65,536
N = 20	No. of levels = 1,048,576

When a voltage amplitude from 0 to V_{max} is used (for example from 0 to 1 Volt), then one quantization step will be:

$$\Delta = V_{max} / \text{No. of levels} = V_{max}/2^N$$

At an adequately high level and complexity of input signal V(t), the digital error (difference between analog signal and stored digital value) from sample to sample will be statistically independent and uniformly distributed in the range of [$-\Delta/2$, $\Delta/2$] where Δ is the step size in the A/D converter. Thus, the maximum Signal-to-Noise Ratio (S/N) in decibels can be calculated to be (Nakajima, 1983; Mieszkowski, 1987):

$$S/N \equiv 20 \cdot \log (V_{SIGNAL\ RMS} / V_{NOISE\ RMS})$$
$$= 6.02 \cdot N + 1.76 \text{ [dB]} \approx 6 \cdot N \text{ [dB]}$$

for all, practical purposes.

Thus, converter resolutions of 8, 12,16, and 20 bits would allow a 48, 72, 96, and 120 dB S/N ratio respetively.

A Digital Recording/Processing System

A block diagram of a digital recording/processing system

is shown in figure. The processes at each of the numbered blocks 1 to 7 are described below:

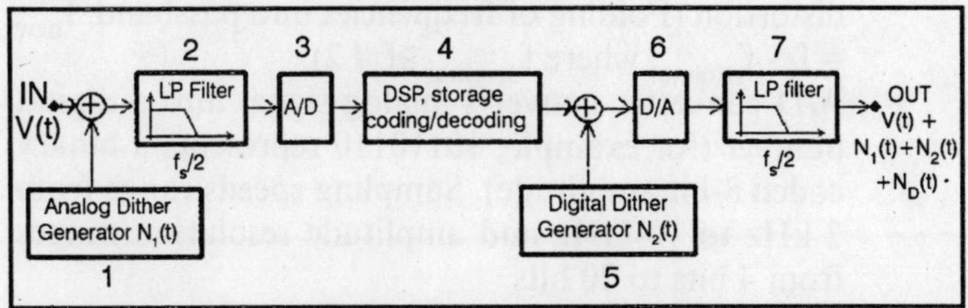

Fig. Block Diagram of Digital Recording Processing System.

Both sources of noise [$N_1(t)$, $N_2(t)$] are needed in order to avoid digital distortions of the signal V(t) in the form of coherent noise $N_D(t)$. Properly chosen $N_1(t)$ and $N_2(t)$ add only a little noise to the output, but remove coherence of $N_D(t)$ (digital noise) with the signal V(t).

- Following Nakajima (1983), Mieszkowski (1989) and Wannamaker, Lipshitz and Vanderkooy (1989), analog dither must be added to the input signal in order to
  - linearize the A/D converter
  - make possible improvement of S/N by averaging process according to formula:
    (S/N) after averaging = (S/N) before averaging $n^{1/2}$ (5)
    where: n = No. of averaged signals
  - eliminate harmonic distortions (created when digital noise $N_D(t)$ is coherent with signal V(t)).
  - eliminate intermodulation distortion (created as well when digital noise $N_D(t)$ is coherent with signal V(t)).
  - eliminate "digital deafness" (when the signal V(t) falls below $\pm\Delta$, where Δ is the step size in the A/D converter, the signal will not be recorded at all unless there is a noise $N_1(t)$ on the input).
  - eliminate noise modulation by the signal
- Input low pass filter (antialiasing filter) should

eliminate all frequencies above $f_s/2$, where f_s = sampling frequency, in order to avoid aliasing distortion (Folding of frequencies into passband: $f_{new} = f_s - f_{original}$ where $f_{original} \geq f_s / 2$).

- A/D converter converts analog signal into a digital number (for example, 10110110 represents a binary coded 8-bit amplitude). Sampling speeds range from 2 kHz to 10 GHz and amplitude resolution ranges from 4 bits to 20 bits.
- If DSP is performed on the signal, one must add digital dither $N_2(t)$ (box 5) to avoid digital distortions and coherent noise N_D (t) on the output of D/A converter. Digital processing should also be performed using sufficiently precise real numbers to avoid round-off errors.

Storage of digital data can be performed on magnetic tape, optical disk, magnetic disk, or RAM (Random Access Memory).

Prior to storage, extra code is generated to allow for error correction.

This error correction code allows detection and *correction* of errors during playback of the audio signal. Redundant information must be added to the original signal in order to combat noise inherent in any storage/communication system.

The particular type of code and error correction system depends on storage medium, communication channel used and immunity from errors (an arbitrarily small probability of error can be obtained, Nakajima, 1983; Shannon, 1949/1975).

- Prior to D/A conversion, digital dither must be added to numbers representing amplitude of the signal if DSP has been performed. Optimal digital dither has triangular probability density function (PDF) (Wannamaker, et al. 1989).
- D/A converter converts digital numbers into analog signal. Available conversion speeds are 2 kHz to 200 MHz and available amplitude resolution is 4 bits to 20 bits.
- Output low pass filter should eliminate all frequencies

above f_s /2 which are generated during D/A conversion.

DIGITAL ADVANTAGES

Table I summarizes the author's comparison of studio quality reel-to-reel analog tape recorder with 16 bit digital recorder. These data are derived from specifications by various manufacturers of analog and digital audio products. This table implies that the digital recorder has many advantages over its analog counterpart Performance of the analog recorder depends very much on the calibration and tape used, as well as on the environmental conditions such as temperature and humidity. This is not the case for a digital recorder, as long as errors generated are within the limits of error correctability of the particular device.

Table. Comparison of Analog reel to Reel Recorder with 16 bit Digital Recorder.

Parameter	Reel to Reel Tape Recorder	16 Bit Digital Recorder
S/N Ratio	65 dB (linear system)	93 dB (linear system)
Total harmonic distortion	0.2%	0.005%
Wow and Flutter	0.03%	UNMEASURABLE (Quartz Accuracy)
Frequency Response	30-20,000 Hz (±3 dB)	0-20,000 Hz (±0.5 dB)
Loss of S/N during copying	3 dB	NONE (PERFECT COPY) - as long as done in digital domain
Deterioration over time	YES	NONE (as long as within limits of correctability of the system)

DIGITAL FORMATS

Common Coding Systems

Below is a short list of commonly used digital coding algorithms (using as an example a single channel digital recording system with swnpling frequency f_s = 44,100 Hz and 16 bit A/D and D/A conversion). The data compression algorithms, which are more efficient than PCM (use less storage space), preserve the information content of the signal.

Not mentioned here are data reduction/compression algorithms, which reduce information content of the original signal (arbitrarily or on the basis of psychoacoustics research results). PCM - PCM was invented by A.H. Reeves in 1939 (American Patents 2272070, 1942-2 see Nakajima, 1983) and was analyzed and developed as a modulation system from the point of view of communication theory by C.E. Shannon (1949). Using only two alternative pulse values (0 and 1), a 16- pulse train is generated which indicates the sampled value (for example, 1010 1111 0110 1101, a binary coded 16 bit number). During conversion, 16 bit amplitudes A1, A2, A3... are generated with a rate 44,100/sec. The demand on the storage device and speed of transmission channel is 88,200 Bytes/sec. This is a 'brute force' approach, which is not the most effective way of using the storage device and transmission channel. DPCM - Differential Pulse Code Modulation.

During conversion only 4 bit (for example) differences between consecutive amplitudes are generated (A2-A1), (A3-A2), (A4-A3)... at the rate of 44,100 /sec. Demand on the storage device and speed of transmission channel is 22,050 Bytes/sec. ADPCM - Adaptive Differential Pulse Code Modulation. Depending on the signal, the number of available bits to represent the difference between consecutive 16 bit samples is varied. For example, for the case of total quiet at the input (or small signal) the difference could be switched off totally or represented only by 1 bit. Demand on the storage device and the speed of transmission channel could vary between 0 Bytes/sec and 88,200 Bytes/sec depending on signal complexity. This is probably the most effective way of coding. Similar means of coding could be used for video signals because there is not much change from frame to frame most of the time. M - Delta Modulation. During coding only 1 bit differences between consecutive amplitudes are generated at a high conversion speed indicating whether the signal was increased or decreased (from the previous sample). Demand on the storage device and the speed of transmission channel is very high in comparison to the PCM system for the same quality of signal (Nakajima, et al., 1983).

Recording/Storage Systems

Listed below are current common recording/storage systems for digital audio data. PCM unit + VCR recorder - 2 and 4 channels. These are professional and semi-professional systems with 14 bit or 16 bit resolution and 44,056 Hz or 44,100 Hz sampling frequency. The PCM signal is stored on video tape in pseudo-video format. Most of the early systems were of this type. DASH (Digital Audio Stationary Head Recorder)

This is a professional 16 bit system with up to 48 tracks. Available are 40,056, 44,100 and 48,000 Hz sampling frequencies. R-DAT (Rotating Head Digital Audio Tape Recoder). This is a professional and consumer 2-channel system with 16 bit resolution and 32,000, 44,056, 44,100, 48,000 Hz sampling frequencies. Magnetic Hard Disk and RAM (Random Access Memory) based Recorders.

These are computer based professional and semi-professional recording systems having from 1 to 24 tracks. The resolution is from 8 to 18 bits. Sampling frequencies are from 2 kHz to 250 kHz. The computers may be common microcomputers as well as main-frame computers. They offer the highest flexibility in terms of digital editing of stored sound and in the author's opinion are the trend of the future.

Optical WMRM (Write Many Read Many), Erasable Optical Disk based Recorders. This format is becoming popular for audio applications because the removable optical cartridge can store about 600 MBytes of data and is more robust than magnetic media. Writing and reading is done by laser without physical contact with the disk. The NeXT computer has the first commercially available optical disk drive with 256 MBytes capacity (Thmpson and Baran, 1988). Also, Nakamichi recently showed during the AES 7th International Conference a working prototype of an optical disk recorder, similar to a CD player (Mascenik, 1989).

Applications

Digital techniques for storing and transmission of audio signals are attractive because they offer high quality signals, which do not deteriorate with transmission distance, number of copies or time. Digital information when properly stored and transmitted maintains its 100% integrity in contrast to analog information which deteriorates during each transmission and storage cycle. DSP is also far more powerful than ASP (Analog Signal Processing). First, the quality of the signal is maintained during DSP.

Second, most of the DSP devices are very flexible because one can run many different applications on the same hardware by a change of the software. Analog devices are devoted to particular tasks and are not as flexible. Third, digital signal processing can perform operations impossible in the analogue domain. Some of the functions which could be performed by the DSP devices are: filtering, equalization, compression/expansion of dynamic range, time compression/expansion, delay, reverberation, pitch change, generation of arbitrary signal or noise, music and voice synthesis, noise reduction, signal restoration, automatic pattern and voice recognition, time- reverse, noise gate, automatic gain control, mixing of signals, and FFT (Fast Fourier Transforms).

In recent years DSP units have become relatively affordable. Also, there are many products available as plug-in cards for popular microcomputers, which contain DSP chips from such manufacturers as Motorola or Texas Instruments. DSP systems based on microcomputers are relatively fast (but not as fast as devoted hardware) and very flexible. The future of digital recording and DSP looks very bright. Higher speeds of microprocessors and DSP chips make real time applications of even complex algorithms realistic. Falling prices of RAM chips and storage devices like the erasable optical disk make them affordable for many researchers and musicians. In the author's opinion it is almost certain that the majority of future recording and DSP equipment will be based on microcomputers. Storage media of the future will probably be erasable optical disks and RAM cards. With falling prices of RAM chips and already available 4 Mbit chips in a single package, one can expect portable RAM based ADPCM recorders to replace mechanically complex R-DAT machines in the near future.

Chapter 7

Discovery of Infrared

Sir Frederick William Herschel (1738-1822) was born in Hanover, Germany, and became well known as both a musician and an astronomer. He emigrated to England in 1757, and with his sister Caroline, constructed telescopes to survey the night sky. Their work resulted in several catalogs of double stars and nebulae. William Herschel is perhaps most famous for his discovery of the planet Uranus in 1781, the first new planet found since antiquity. Caroline Herschel gained fame for the discovery of several comets.

In the year 1800, William Herschel made another very important discovery. He was interested in learning how much heat passed through the different colored filters he used to observe the Sun and noticed that filters of different colors seemed to pass different levels of heat. Herschel thought that the colors themselves might contain different levels of heat, so he devised a clever experiment to investigate his hypothesis.

Herschel directed sunlight through a glass prism to create a spectrum - the "rainbow" created when light is divided into its colors - and measured the temperature of each colour. He used three thermometers with blackened bulbs (to better absorb the heat) and placed one bulb in each colour while the other two were placed beyond the spectrum as control samples. As he measured the temperatures of the violet, blue, green, yellow, orange and red light, he noticed that all of the colors had temperatures higher than the controls and that the temperature of the colors increased from the violet to the red part of the spectrum. After noticing this pattern, Herschel decided to measure the temperature just *beyond* the red portion of the spectrum in a region apparently devoid of sunlight. To his surprise, he found that this region had the highest temperature of all.

Herschel performed further experiments on what he called the "calorific rays" that existed beyond the red part of the spectrum and found that they were reflected, refracted, absorbed and transmitted just like visible light. What Sir William had discovered was a form of light (or radiation) beyond red light. These "calorific rays" were later renamed infrared rays or infrared radiation (the prefix infra means `below'). Herschel's experiment was important not only because it led to the discovery of infrared, but also because it was the first time that someone showed that there were forms of light that we cannot see with our eyes. Herschel's original prism and mirror are on display at the National Museum of Science and Industry in London, England.

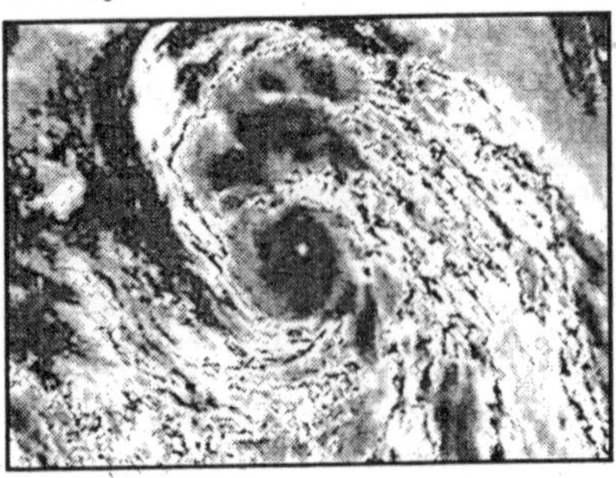

Fig. Infrared Image of a Hurricane (NASA)

Today,infrared technology has many exciting and useful applications. In the field of infrared astronomy, new and fascinating discoveries are being made about the Universe. Medical infrared imaging is a very useful diagnostic tool. Infrared cameras are used for police and security work as well as in fire fighting and in the military. Infrared imaging is used to detect heat loss in buildings and in testing electronic systems. Infrared satellites have been used to monitor the Earth's weather, to study vegetation patterns, and to study geology and ocean temperatures

INFRARED

Our eyes are detectors which are designed to detect visible light waves (or visible radiation). Visible light is one of the few types of radiation that can penetrate our atmosphere and be detected on the Earth's surface. As we have seen from the section on the discovery of infrared, there are forms of light (or radiation) which we cannot see. Actually we can only see a very small part of the entire range of radiation called the electromagnetic spectrum.

The electromagnetic spectrum includes gamma rays, X-rays, ultraviolet, visible, infrared, microwaves, and radio waves. The only difference between these different types of radiation is their wavelength or frequency. Wavelength increases and frequency (as well as energy and temperature) decreases from gamma rays to radio waves. All of these forms of radiation travel at the speed of light (186,000 miles or 300,000,000 meters per second in a vacuum). In addition to visible light, radio, some infrared and a very small amount of ultraviolet radiation also reaches the Earth's surface from space. Fortunately for us, our atmosphere blocks out the rest, much of which is very hazardous, if not deadly, for life on Earth.

Infrared radiation lies between the visible and microwave portions of the electromagnetic spectrum. Infrared waves have wavelengths longer than visible and shorter than microwaves, and have frequencies which are lower than visible and higher than microwaves.

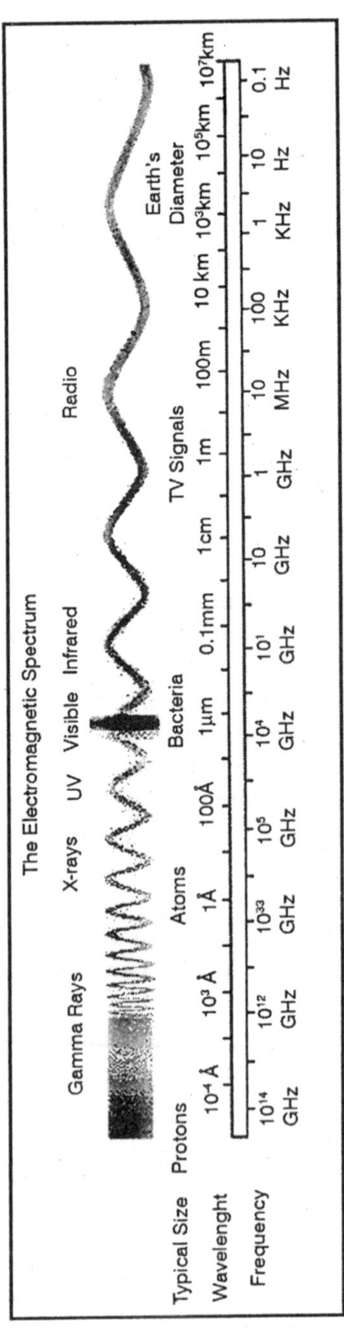
The Electromagnetic Spectrum
Gamma Rays
X-rays
UV
Visible
Infrared
Radio
Typical Size
Wavelenght
Frequency
Protons
Atoms
Bacteria
TV Signals
Earth's Diameter
10^{-4} Å
10^{3} Å
1Å
100Å
1µm
0.1mm
1cm
1m
100m
10 km
10^{3}km
10^{5}km
10^{7}km
10^{14} GHz
10^{12} GHz
10^{33} GHz
10^{5} GHz
10^{4} GHz
10^{1} GHz
10 GHz
1 GHz
10 MHz
100 KHz
1 KHz
10 Hz
0.1 Hz

Infrared is broken into three categories: near, mid and far-infrared. Near-infrared refers to the part of the infrared spectrum that is closest to visible light and far-infrared refers to the part that is closer to the microwave region. Mid-infrared is the region between these two.

Fig. Courtesy of Inframetrics

The primary source of infrared radiation is heat or thermal radiation. This is the radiation produced by the motion of atoms and molecules in an object. The higher the temperature, the more the atoms and molecules move and the more infrared radiation they produce. Any object which has a temperature i.e. anything above absolute zero (-459.67 degrees Fahrenheit or -273.15 degrees Celsius or 0 degrees Kelvin), radiates in the infrared.

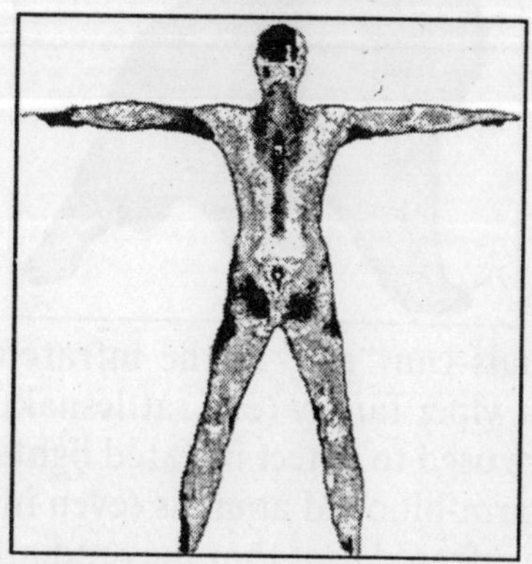

Fig. Courtesy of Meditherm

Absolute zero is the temperature at which all atomic and molecular motion ceases. Even objects that we think of as being very cold, such as an ice cube, emit infrared. When an object is not quite hot enough to radiate visible light, it will emit most of its energy in the infrared. For example, hot charcoal may not give off light but it does emit infrared radiation which we feel as heat. The warmer the object, the more infrared radiation it emits. The infrared image of a landing space shuttle (left) shows the how the tiles underneath the shuttle have been heated during re-entry.

Humans, at normal body temperature, radiate most strongly in the infrared, at a wavelength of about 10 microns (A micron is the term commonly used in astronomy for a micrometer or one millionth of a meter). In the image to the left, the red areas are the warmest, followed by yellow, green and blue (coolest). The image to the right shows a cat in the infrared. The yellow-white areas are the warmest and the purple areas are the coldest. This image gives us a different view of a familiar animal as well as information that we could not get from a visible light picture. Notice the cold nose and the heat from the cat's eyes, mouth and ears.

Some animals can "see" in the infrared. For example, snakes in the pit viper family (e.g. rattlesnakes) have sensory "pits," which are used to detect infrared light. This allows the snake to find warm-blooded animals (even in dark burrows), by detecting the infrared heat that they radiate. Snakes with 2 sensory pits are thought to have some depth perception in the infrared. We experience infrared radiation every day. The heat that we feel from sunlight, a fire, a radiator or a warm sidewalk is infrared. Although our eyes cannot see it, the nerves in our skin can feel it as heat. The temperature-sensitive nerve endings in your skin can detect the difference between your inside body temperature and your outside skin temperature. We also commonly use infrared rays when we operate a television remote.

To learn more about the everyday applications of infrared imaging see our module called "Seeing Our World in a Different Light". We also several new sites showing spectacular infrared images: "Infrared Yellowstone" - infrared images of the geysers, hot springs and mudpots of Yellowstone National Park, The Infrared Zoo, - a wonderful infrared gallery of animals, Infrared Portrait Gallery - an infrared gallery of dozens of people, and our latest Infrared Image Gallery - a collection of numerous infrared images.

IR Astronomy

Infrared Astronomy is the detection and study of the infrared radiation (heat energy) emitted from objects in the Universe. All objects emit infrared radiation. So, Infrared Astronomy involves the study of just about everything in the Universe. In the field of astronomy, the infrared region lies within the range of sensitivity of infrared detectors, which is between wavelengths of about 1 and 300 microns (a micron is one millionth of a meter). The human eye detects only 1% of light at 0.69 microns, and 0.01% at 0.75 microns, and so effectively cannot see wavelengths longer than about 0.75 microns unless the light source is extremely bright.

VIEWING THE INVISIBLE

The Universe sends us a tremendous amount of information in the form of electromagnetic radiation (or light). Much of this information is in the infrared, which we cannot see with our eyes or with visible light telescopes. Only a small amount of this infrared information reaches the Earth's surface, yet by studying this small range of infrared wavelengths, astronomers have uncovered a wealth of new information. Only since the early 1980's have we been able to send infrared telescopes into orbit around the Earth, above the atmosphere which hides most of the Universe's light from us. The new discoveries made by these infrared satellite missions has been astounding. The first of these satellites - IRAS (Infrared Astronomical Satellite) - detected about 350,000 infrared sources, increasing the number of cataloged astronomical sources by about 70%.

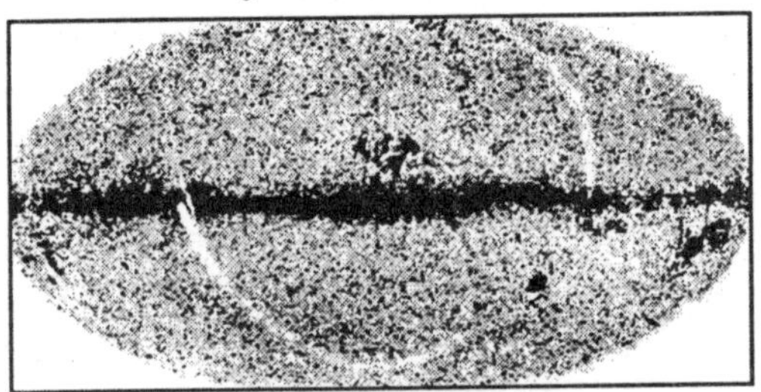

EXPLORING THE HIDDEN UNIVERSE

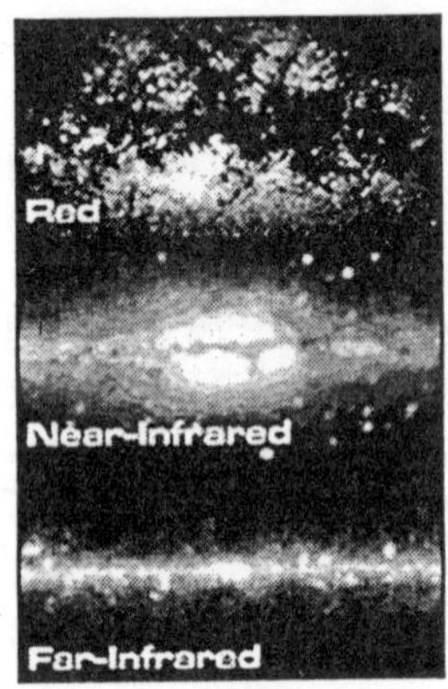

Fig. Galactic Centre

In space, there are many regions which are hidden from optical telescopes because they are embedded in dense regions of gas and dust. However, infrared radiation, having wavelengths which are much longer than visible light, can pass through dusty regions of space without being scattered. This means that we can study objects hidden by gas and dust in the infrared, which we cannot see in visible light, such as the centre of our galaxy and regions of newly forming stars.

The images to the left, of the central region of our own Milky Way Galaxy and of the Cygnus star-forming region, show how areas which cannot be seen in visible light can show up very brightly in the infrared. The top row shows these regions in visible red light. At this wavelength we are seeing the light from billions of stars, particularly the largest, brightest ones. Note the dark bands where vast clouds of dust block our view of more distant objects.The middle row shows the same regions in the near-infrared (infrared wavelengths closest to visible light). The middle row shows the same regions in the near-infrared (infrared wavelengths closest to visible light).

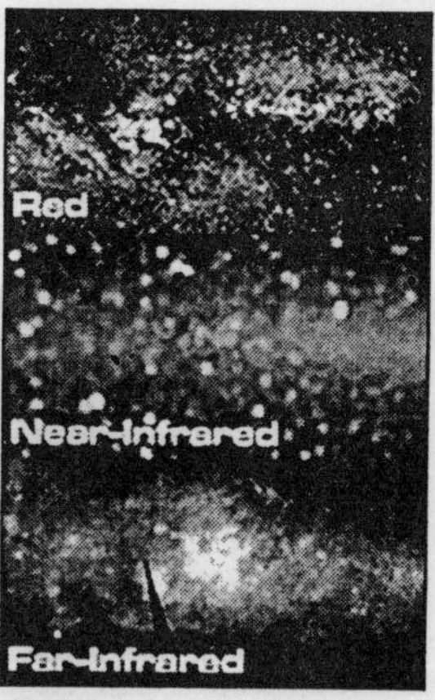

Fig. Cygnus Region

Here the light we see is also generated by stars, but now it better traces the smaller, cooler ones. Notice how the the lanes of dust have become partially transparent, allowing us to see things that are hidden in visible light. Our view of the central bulge of stars in our own Milky Way galaxy is particularly striking since it is almost completely obscured at shorter wavelengths! The bottom images show these regions in the far-infrared (infrared wavelengths farther from visible light). At these wavelengths, stars hardly emit any light at all. Instead almost everything we see is generated by the dust clouds themselves. The dust, which is colder than the coldest arctic night on earth, is still warm enough to emit the thermal infrared radiation seen here.

DETECTING COOL OBJECTS

Many objects in the universe which are much too cool and faint to be detected in visible light, can be detected in the infrared. These include cool stars, infrared galaxies, clouds of particles around stars, nebulae, interstellar molecules, brown dwarfs and planets. For example, the visible light from a planet is hidden by the brightness of the star that it orbits. In the infrared, where planets have their peak brightness, the brightness of the star is reduced, making it possible to detect a planet in the infrared. Some of the most exciting discoveries in infrared astronomy have been the detection of disks of material and possible planets around other stars. Recently, an infrared survey of the Trapezium star cluster in the Orion Nebula revealed over 100 low mass objects which are brown dwarf candidates. Click on the image for details.

EXPLORING THE EARLY UNIVERSE

In the infrared, astronomers can gather information about the universe as it was a very long time ago and study the early evolution of galaxies. As a result of the Big Bang (the tremendous explosion which marked the beginning of our Universe), the universe is expanding and most of the galaxies within it are moving away from each other. Astronomers have discovered that all distant galaxies are moving away from us and that the farther away they are, the faster they are moving. This recession of galaxies away from us has an interesting effect on the light emitted from these galaxies. When an object is moving away from us, the light that it emits is "redshifted".

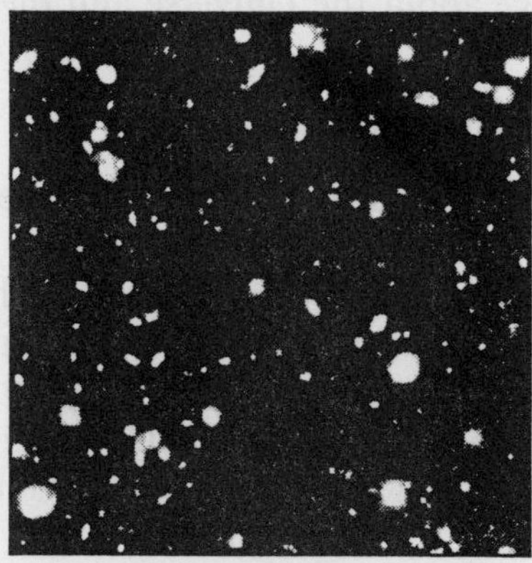

This means that the wavelengths get longer and thereby shifted towards the red part of the spectrum. This effect, called the Doppler effect, is similar to what happens to sound waves emitted from a moving object. For example, if you are standing next to a railroad track and a train passes you while blowing its horn, you will hear the sound change from a higher to a lower frequency as the train passes you by. As a result of this Doppler effect, at large redshifts, all of the ultraviolet and much of the visible light from distant sources is shifted into the infrared part of the spectrum by the time it reaches our telescopes. This means that the only way to study this light is in the infrared. Infrared astronomy will provide a great deal of information on how and when the universe was formed and on what the early universe was like. The image to the left is an infrared view of some of the farthest galaxies ever seen. (Image credit: R.I. Thompson (U. Arizona), NICMOS, HST, NASA)

ADDING TO OUR KNOWLEDGE OF VISIBLE OBJECTS

Objects which can be seen in visible light can also be studied in the infrared. Infrared astronomy can not only allow us to discover new objects and view previously unseen areas of the universe, but it can add to what we already know about visible objects. To get a complete picture of any object in the Universe we need to study all of the radiation that it emits. Infrared Astronomy has, and will continue to, add a great deal to our knowledge about the Universe and the origins of our Solar System.

IR Atmospheric Windows

The Universe sends us light at all wavelengths of the electromagnetic spectrum. However, most of this light does not reach us at ground level here on Earth. Why? Because we have an atmosphere which blocks out many types of radiation while letting other types through. Fortunately for life on Earth, our atmosphere blocks out harmful, high energy radiation like X-rays, gamma rays and most of the ultraviolet rays. It also block out most infrared radiation, as well as very low energy radio waves. On the other hand, our atmosphere lets visible light, most radio waves, and small wavelength ranges of infrared light through, allowing astronomers to view the Universe at these wavelengths.

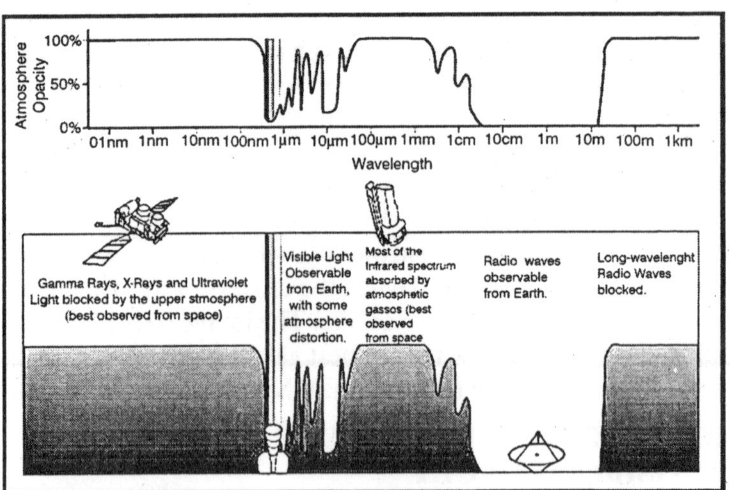

Most of the infrared light coming to us from the Universe is absorbed by water vapour and carbon dioxide in the Earth's atmosphere. Only in a few narrow wavelength ranges, can infrared light make it through (at least partially) to a ground based infrared telescope.

The Earth's atmosphere causes another problem for infrared astronomers. The atmosphere itself radiates strongly in the infrared, often putting out more infrared light than the object in space being observed. This atmospheric infrared emission peaks at a wavelength of about 10 microns (micron is short for a micrometer or one millionth of a meter).

So the best view of the infrared universe, from ground based telescopes, are at infrared wavelengths which can pass through the Earth's atmosphere and at which the atmosphere is dim in the infrared. Ground based infrared observatories are usually placed near the summit of high, dry mountains to get above as much of the atmosphere as possible. Even so, most infrared wavelengths are completely absorbed by the atmosphere and never make it to the ground.

From the table below, you can see that only a few of the infrared "windows" have both high sky transparency and low sky emission. These infrared windows are mainly at infrared wavelengths below 4 microns.

Infrared Windows in the Atmosphere

Wavelength Range	Band	Sky Transparency	Sky Brightness
1.1 - 1.4 microns	J	high	low at night
1.5 - 1.8 microns	H	high	very low
2.0 - 2.4 microns	K	high	very low
3.0 - 4.0 microns	L	3.0 - 3.5 microns: fair 3.5 - 4.0 microns: high	low
4.6 - 5.0 microns	M	low	high
7.5 - 14.5 microns	N	8 - 9 microns and 10 -12 microns: fair others: low	very high
17 - 40 microns	17 - 25 microns: Q 28 - 40 microns: Z	very low	very high
330 - 370 microns		very low	low

Basically, everything we have learned about the Universe comes from studying the light or electromagnetic radiation emitted by objects in space. To get a complete picture of the Universe, we need to see it in all of its light, at all wavelengths. This is why it is so important to send observatories into space, to get above our atmosphere which prevents so much of this valuable information from reaching us.

Since most infrared light is blocked by our atmosphere, infrared astronomers have placed instruments onboard, rockets, balloons, aircraft and space telescopes to view regions of the infrared which are not detectable from the ground. As a result, amazing discoveries about our Universe have been made and hundreds of thousands of new astronomical sources have been detected for the first time.

Due to the rapid development of better infrared detectors and the ability to place telescopes in space, the future is extremely bright for infrared astronomy.

Ground based infrared observatories, using advanced techniques such as Adaptive Optics are providing fascinating views of the infrared Universe viewed through our atmosphere's infrared windows.

Fig. Mauna Kea Observatories

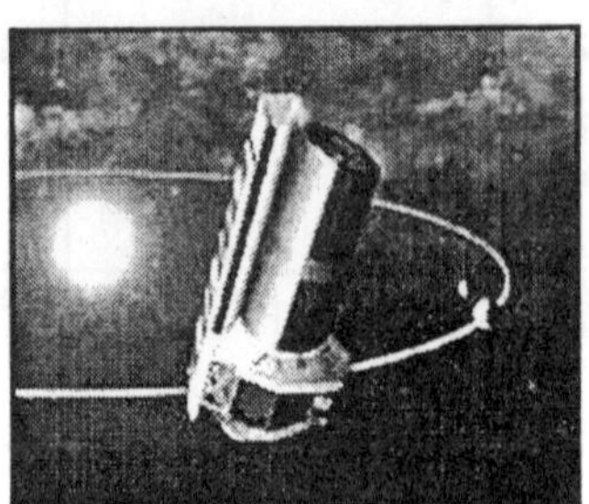

Although these observatories cannot view at other infrared wavelengths, they can observe the near-infrared sky almost anytime the weather permits, providing valuable long term studies of objects in space.

New missions are being planned to get above the atmosphere to observe the infrared Universe with better resolution than ever before. SOFIA, an airborne observatory, is schedule to start operations in 2004. The Spitzer Space Telescope, launched in August 2003, is NASA's next great observatory in space.

In the next decade, you will probably hear much news about discoveries being made in infrared astronomy, as we now can see beyond our atmosphere's infrared windows!

Near, Mid and Far Infrared

Infrared is usually divided into 3 spectral regions: near, mid and far-infrared. The boundaries between the near, mid and far-infrared regions are not agreed upon and can vary. The main factor that determines which wavelengths are included in each of these three infrared regions is the type of detector technology used for gathering infrared light.

Near-infrared observations have been made from ground based observatories since the 1960's. They are done in much the same way as visible light observations for wavelengths less than 1 micron, but require special infrared detectors beyond 1 micron. Mid and far-infrared observations can only be made by observatories which can get above our atmosphere. These observations require the use of special cooled detectors containing crystals like germanium whose electrical resistance is very sensitive to heat.

Infrared radiation is emitted by any object that has a temperature (ie radiates heat). So, basically all celestial objects emit some infrared. The wavelength at which an object radiates most intensely depends on its temperature. In general, as the temperature of an object cools, it shows up more prominently at farther infrared wavelengths. This means that some infrared wavelengths are better suited for studying certain objects than others.

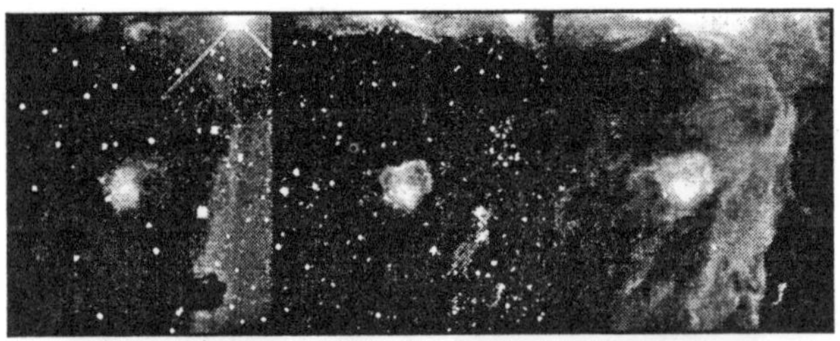

Fig. Visible (courtesy of Howard McCallon), near-infrared (2MASS), and mid-infrared (ISO) view of the Horsehead Nebula. Image assembled by Robert Hurt.

Spectral region	Wavelength range (microns)	Temperature range (degrees kelvin)	What we see
Near-Infrared	(0.7-1) to 5	740 to (3,000-5,200)	Cooler red stars Red giants Dust is transparent
Mid-Infrared	5 to (25-40)	(92.5-140) to 740	Planets, comets and asteroids Dust warmed by starlight Protoplanetary disks
Far-Infrared	(25-40) to (200-350)	(10.6-18.5) to (92.5-140)	Emission from cold dust Central regions of galaxies Very cold molecular clouds

As we move from the near-infrared into mid and far-infrared regions of the spectrum, some celestial objects will appear while others will disappear from view. For example, in the above image you can see how more stars (generally cooler stars) appear as we go from the visible light image to the near-infrared image. In the near-infrared, the dust also becomes transparent, allowing us to see regions hidden by dust in the visible image. As we go to the mid-infrared image, the cooler dust itself glows. The table below highlights what we see in the different infrared spectral regions.

NEAR INFRARED

Between about 0.7 to 1.1 microns we can use the same observing methods as are use for visible light observations, except for observation by eye. The infrared light that we observe in this region is not thermal (not due to heat radiation). Many do not even consider this range as part of infrared astronomy. Beyond about 1.1 microns, infrared emission is primarily heat or thermal radiation.

As we move away from visible light towards longer wavelengths of light, we enter the infrared region. As we enter the near-infrared region, the hot blue stars seen clearly in visible light fade out and cooler stars come into view. Large red giant stars and low mass red dwarfs dominate in the near-infrared. The near-infrared is also the region where interstellar dust is the most transparent to infrared light.

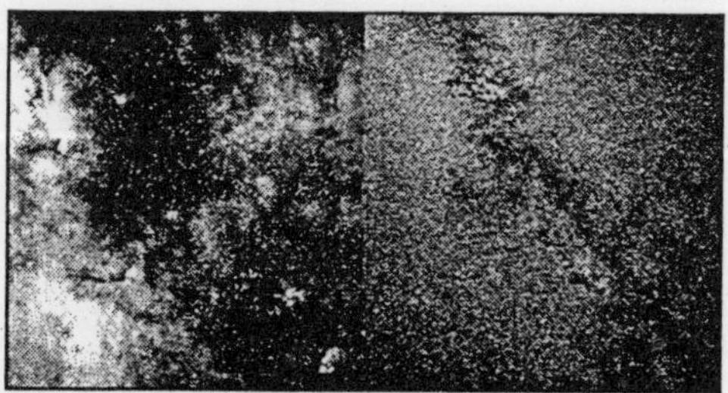

Fig. Visible (left) and Near-Infrared View of the Galactic Centre

Notice in the above images how centre of our galaxy, which is hidden by thick dust in visible light (left), becomes transparent in the near-infrared (right). Many of the hotter stars in the visible image have faded in the near-infrared image.

The near-infrared image shows cooler, reddish stars which do not appear in the visible light view. These stars are primarily red dwarfs and red giants.

Red giants are large reddish or orange stars which are running out of their nuclear fuel. They can swell up to 100 times their original size and have temperatures which range from 2000 to 3500 K. Red giants radiate most intensely in the near-infrared region.

Red dwarfs are the most common of all stars. They are much smaller than our Sun and are the coolest of the stars having a temperature of about 3000 K which means that these stars radiate most strongly in the near-infrared. Many of these stars are too faint in visible light to even be detected by optical telescopes, and have been discovered for the first time in the near-infrared.

MID INFRARED

As we enter the mid-infrared region of the spectrum, the cool stars begin to fade out and cooler objects such as planets, comets and asteroids come into view. Planets absorb light from the sun and heat up. They then re-radiate this heat as infrared light. This is different from the visible light that we see from the planets which is reflected sunlight. The planets in our solar system have temperatures ranging from about 53 to 573 degrees Kelvin.

Fig. An infrared View of the Earth

Objects in this temperature range emit most of their light in the mid-infrared. For example, the Earth itself radiates most strongly at about 10 microns. Asteroids also emit most of their light in the mid-infrared making this wavelength band the most efficient for locating dark asteroids. Infrared data can help to determine the surface composition, and diameter of asteroids.

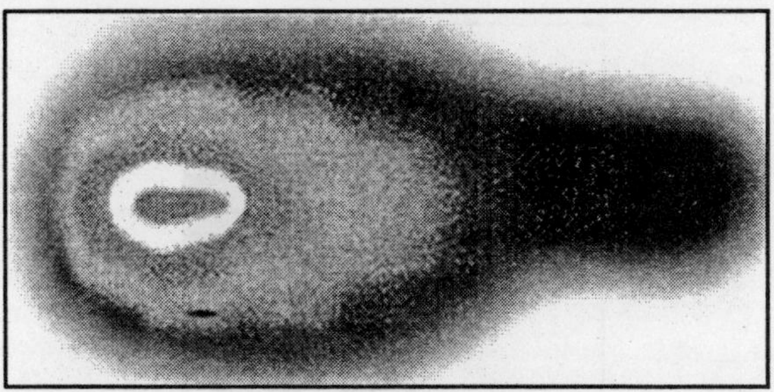

Fig. IRAS mid-infrared View of Comet IRAS-Āraki-Alcock

Dust warmed by starlight is also very prominent in the mid-infrared. An example is the zodiacal dust which lies in the plane of our solar system. This dust is made up of silicates (like the rocks on Earth) and range in size from a tenth of a micron up to the size of large rocks. Silicates emit most of their radiation at about 10 microns. Mapping the distribution of this dust can provide clues about the formation of our own solar system. The dust from comets also has strong emission in the mid-infrared.

Warm interstellar dust also starts to shine as we enter the mid-infrared region. The dust around stars which have ejected material shines most brightly in the mid-infrared. Sometimes this dust is so thick that the star hardly shines through at all and can only be detected in the infrared.

Sometimes this dust is so thick that the star hardly shines through at all and can only be detected in the infrared. Protoplanetary disks, the disks of material which surround newly forming stars, also shines brightly in the mid-infrared. These disks are where new planets are possibly being formed.

FAR INFRARED

In the far-infrared, the stars have all vanished. Instead we now see very cold matter (140 Kelvin or less). Huge, cold clouds of gas and dust in our own galaxy, as well as in nearby galaxies, glow in far-infrared light. In some of these clouds, new stars are just beginning to form. Far-infrared observations can detect these protostars long before they "turn on" visibly by sensing the heat they radiate as they contract."

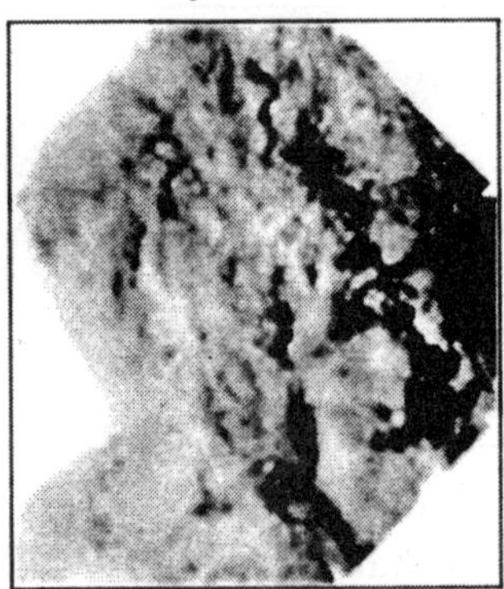

Fig. IRAS View of Infrared Cirrus - Dust Heated by Starlight

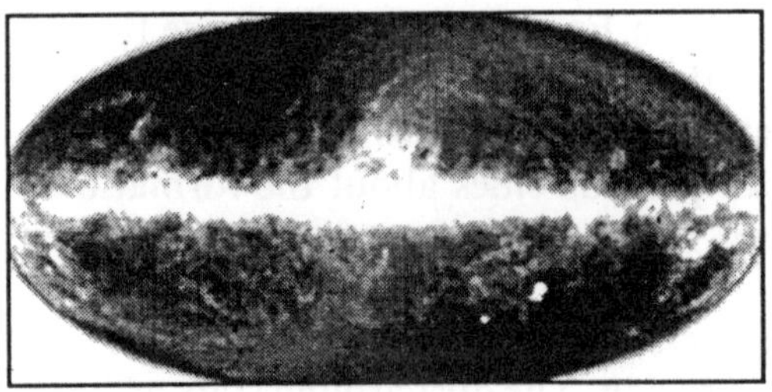

Fig. Michael Hauser (Space Telescope Science Institute), the COBE/DIRBE Science Team, and NASA

The centre of our galaxy also shines brightly in the far-infrared because of the thick concentration of stars embedded in dense clouds of dust. These stars heat up the dust and cause it to glow brightly in the infrared. The image (at left) of our galaxy taken by the COBE satellite, is a composite of far-infrared wavelengths of 60, 100, and 240 microns.

Except for the plane of our own Galaxy, the brightest far-infrared object in the sky is central region of a galaxy called M82. The nucleus of M82 radiates as much energy in the far-infrared as all of the stars in our Galaxy combined. This far-infrared energy comes from dust heated by a source that is hidden from view. The central regions of most galaxies shine very brightly in the far-infrared. Several galaxies have active nuclei hidden in dense regions of dust. Others, called starburst galaxies, have an extremely high number of newly forming stars heating interstellar dust clouds. These galaxies, far outshine all others galaxies in the far-infrared.

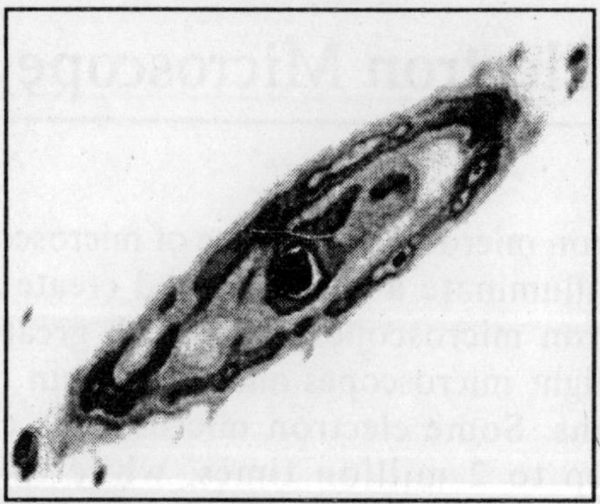

Fig. IRAS Infrared View of the Andromeda Galaxy (M31) - Notice the Bright Central Region.

Chapter 8

Electron Microscope

An electron microscope is a type of microscope that uses electrons to illuminate a specimen and create an enlarged image. Electron microscopes have much greater resolving power than light microscopes and can obtain much higher magnifications. Some electron microscopes can magnify specimens up to 2 million times, while the best light microscopes are limited to magnifications of 2000 times. Both electron and light microscopes have resolution limitations, imposed by their wavelength. The greater resolution and magnification of the electron microscope is due to the wavelength of an electron, its de Broglie wavelength, being much smaller than that of a light photon, electromagnetic radiation.

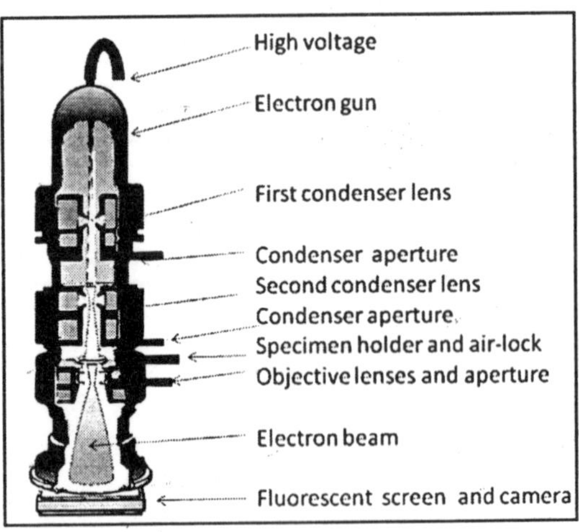

Fig. Diagram of a Transmission Electron Microscope

The electron microscope uses electrostatic and electromagnetic lenses in forming the image by controlling the electron beam to focus it at a specific plane relative to the specimen in a manner similar to how a light microscope uses glass lenses to focus light on or through a specimen to form an image.

The first electron microscope prototype was built in 1931 by the German engineers Ernst Ruska and Max Knoll. Although this initial instrument was capable of magnifying objects by only four hundred times, it demonstrated the principles of an electron microscope. Two years later, Ruska constructed an electron microscope that exceeded the resolution possible with an optical microscope.

Reinhold Rudenberg, the scientific director of Siemens, had patented the electron microscope in 1931, stimulated by family illness to make the poliomyelitis virus particle visible. In 1937 Siemens began funding Ruska and Bodo von Borries to develop an electron microscope. Siemens also employed Ruska's brother Helmut to work on applications, particularly with biological specimens.

In the same decade Manfred von Ardenne pioneered the scanning electron microscope and his universal electron microscope.

Siemens produced the first commercial TEM in 1939, but the first practical electron microscope had been built at the University of Toronto in 1938, by Eli Franklin Burton and students Cecil Hall, James Hillier, and Albert Prebus.

Although modern electron microscopes can magnify objects up to two million times, they are still based upon Ruska's prototype. The electron microscope is an essential item of equipment in many laboratories. Researchers use them to examine biological materials (such as microorganisms and cells), a variety of large molecules, medical biopsy samples, metals and crystalline structures and the characteristics of various surfaces. The electron microscope is also used extensively for inspection, quality assurance and failure analysis applications in industry, including, in particular, semiconductor device fabrication.

TYPES OF ELECTRON MICROSCOPE

TRANSMISSION ELECTRON MICROSCOPE (TEM)

The original form of electron microscope, the transmission electron microscope (TEM) uses a high voltage electron beam to create an image. The electrons are emitted by an electron gun, commonly fitted with a tungsten filament cathode as the electron source. The electron beam is accelerated by an anode typically at +100keV (40 to 400 keV) with respect to the cathode, focused by electrostatic and electromagnetic lenses, and transmitted through the specimen that is in part transparent to electrons and in part scatters them out of the beam.

When it emerges from the specimen, the electron beam carries information about the structure of the specimen that is magnified by the objective lens system of the microscope. The spatial variation in this information (the "image") is viewed by projecting the magnified electron image onto a fluorescent viewing screen coated with a phosphor or scintillator material such as zinc sulfide. The image can be photographically recorded by exposing a photographic film or plate directly to the electron beam, or a high-resolution phosphor may be coupled by means of a lens optical system or a fibre optic light-guide to the sensor of a CCD (charge-coupled device) camera. The image detected by the CCD may be displayed on a monitor or computer.

Resolution of the TEM is limited primarily by spherical aberration, but a new generation of aberration correctors have been able to partially overcome spherical aberration to increase resolution. Software correction of spherical aberration for the High Resolution TEM (HRTEM) has allowed the production of images with sufficient resolution to show carbon atoms in diamond separated by only 0.89 εngstrφm (89 picometers) and atoms in silicon at 0.78 εngstrφm (78 picometers) at magnifications of 50 million times. The ability to determine the positions of atoms within materials has made the HRTEM an important tool for nano-technologies research and development.

Unlike the TEM, where electrons of the high voltage beam carry the image of the specimen, the electron beam of the Scanning Electron Microscope (SEM) does not at any time carry a complete image of the specimen.

The SEM produces images by probing the specimen with a focused electron beam that is scanned across a rectangular area of the specimen. At each point on the specimen the incident electron beam loses some energy, and that lost energy is converted into other forms, such as heat, emission of low-energy secondary electrons, light emission (cathodoluminescence) or x-ray emission.

The display of the SEM maps the varying intensity of any of these signals into the image in a position corresponding to the position of the beam on the specimen when the signal was generated.

In the SEM image of an ant shown at right, the image was constructed from signals produced by a secondary electron detector, the normal or conventional imaging mode in most SEMs.

Generally, the image resolution of an SEM is about an order of magnitude poorer than that of a TEM. However, because the SEM image relies on surface processes rather than transmission it is able to image bulk samples up to several centimetres in size (depending on instrument design) and has a much greater depth of view, and so can produce images that are a good representation of the 3D structure of the sample.

REFLECTION ELECTRON MICROSCOPE (REM)

In the Reflection Electron Microscope (REM) as in the TEM, an electron beam is incident on a surface, but instead of using the transmission (TEM) or secondary electrons (SEM), the reflected beam of elastically scattered electrons is detected.

This technique is typically coupled with Reflection High Energy Electron Diffraction and *Reflection high-energy loss spectrum (RHELS)*. Another variation is Spin-Polarized Low-Energy Electron Microscopy (SPLEEM), which is used for looking at the microstructure of magnetic domains.

SCANNING TRANSMISSION ELECTRON MICROSCOPE (STEM)

The STEM rasters a focused incident probe across a specimen that (as with the TEM) has been thinned to facilitate detection of electrons scattered *through* the specimen. The high resolution of the TEM is thus possible in STEM. The focusing action (and aberrations) occur before the electrons hit the specimen in the STEM, but afterward in the TEM. The STEM's use of SEM-like beam rastering simplifies annular dark-field imaging, and other analytical techniques, but also means that image data is acquired in serial rather than in parallel fashion.

Sample Preparation

Materials to be viewed under an electron microscope may require processing to produce a suitable sample. The technique required varies depending on the specimen and the analysis required:

- Chemical Fixation for biological specimens aims to stabilize the specimen's mobile macromolecular structure by chemical crosslinking of proteins with aldehydes such as formaldehyde and glutaraldehyde, and lipids with osmium tetroxide.
- *Cryofixation* – freezing a specimen so rapidly, to liquid nitrogen or even liquid helium temperatures, that the water forms vitreous (non-crystalline) ice. This preserves the specimen in a snapshot of its solution state. An entire field called cryo-electron microscopy has branched from this technique. With the development of cryo-electron microscopy of vitreous sections (CEMOVIS), it is now possible to observe virtually any biological specimen close to its native state.[citation needed]
- *Dehydration* – freeze drying, or replacement of water with organic solvents such as ethanol or acetone, followed by critical point drying or infiltration with embedding resins.
- *Embedding, biological specimens* – after dehydration,

tissue for observation in the transmission electron microscope is embedded so it can be sectioned ready for viewing. To do this the tissue is passed through a 'transition solvent' such as epoxy propane and then infiltrated with a resin such as Araldite epoxy resin; tissues may also be embedded directly in water-miscible acrylic resin. After the resin has been polymerised (hardened) the sample is thin sectioned (ultrathin sections) and stained - it is then ready for viewing.

- *Embedding, materials* - after embedding in resin, the specimen is usually ground and polished to a mirror-like finish using ultra-fine abrasives. The polishing process must be performed carefully to minimise scratches and other polishing artefacts that reduce image quality.
- *Sectioning* – produces thin slices of specimen, semitransparent to electrons. These can be cut on an ultramicrotome with a diamond knife to produce ultrathin slices about 60-90nm thick. Disposable glass knives are also used because they can be made in the lab and are much cheaper.
- *Staining* – uses heavy metals such as lead, uranium or tungsten to scatter imaging electrons and thus give contrast between different structures, since many (especially biological) materials are nearly "transparent" to electrons (weak phase objects). In biology, specimens are can be stained "en bloc" before embedding and also later after sectioning. Typically thin sections are stained for several minutes with an aqueous or acoholic solution of uranyl acetate followed by aqueous lead citrate.
- *Freeze-fracture or freeze-etch* – a preparation method particularly useful for examining lipid membranes and their incorporated proteins in "face on" view. The fresh tissue or cell suspension is frozen rapidly (cryofixed), then fractured by simply breaking or by using a microtome while maintained at liquid

nitrogen temperature. The cold fractured surface (sometimes "etched" by increasing the temperature to about -100°C for several minutes to let some ice sublime) is then shadowed with evaporated platinum or gold at an average angle of 45° in a high vacuum evaporator. A second coat of carbon, evaporated perpendicular to the average surface plane is often performed to improve stability of the replica coating. The specimen is returned to room temperature and pressure, then the extremely fragile "pre-shadowed" metal replica of the fracture surface is released from the underlying biological material by careful chemical digestion with acids, hypochlorite solution or SDS detergent. The still-floating replica is thoroughly washed from residual chemicals, carefully fished up on EM grids, dried then viewed in the TEM.

- *Ion Beam Milling* – thins samples until they are transparent to electrons by firing ions (typically argon) at the surface from an angle and sputtering material from the surface. A subclass of this is Focused ion beam milling, where gallium ions are used to produce an electron transparent membrane in a specific region of the sample, for example through a device within a microprocessor. Ion beam milling may also be used for cross-section polishing prior to SEM analysis of materials that are difficult to prepare using mechanical polishing.
- *Conductive Coating* – An ultrathin coating of electrically-conducting material, deposited either by high vacuum evaporation or by low vacuum sputter coating of the sample. This is done to prevent the accumulation of static electric fields at the specimen due to the electron irradiation required during imaging. Such coatings include gold, gold/palladium, platinum, tungsten, graphite etc. and are especially important for the study of specimens with the scanning electron microscope. Another reason for

coating, even when there is more than enough conductivity, is to improve contrast, a situation more common with the operation of a FESEM (field emission SEM). When an osmium coater is used, a layer far thinner than would be possible with any of the previously mentioned sputtered coatings is possible.

Disadvantages

Electron microscopes are expensive to build and maintain, but the capital and running costs of confocal light microscope systems now overlaps with those of basic electron microscopes. They are dynamic rather than static in their operation, requiring extremely stable high-voltage supplies, extremely stable currents to each electromagnetic coil/lens, continuously-pumped high- or ultra-high-vacuum systems, and a cooling water supply circulation through the lenses and pumps. As they are very sensitive to vibration and external magnetic fields, microscopes designed to achieve high resolutions must be housed in stable buildings (sometimes underground) with special services such as magnetic field cancelling systems. Some desktop low voltage electron microscopes have TEM capabilities at very low voltages (around 5 kV) without stringent voltage supply, lens coil current, cooling water or vibration isolation requirements and as such are much less expensive to buy and far easier to install and maintain, but do not have the same ultra-high (atomic scale) resolution capabilities as the larger instruments.

The samples largely have to be viewed in vacuum, as the molecules that make up air would scatter the electrons. One exception is the environmental scanning electron microscope, which allows hydrated samples to be viewed in a low-pressure (up to 20Torr/2.7kPa), wet environment.

Scanning electron microscopes usually image conductive or semi-conductive materials best. Non-conductive materials can be imaged by an environmental scanning electron microscope. A common preparation technique is to coat the sample with a several-nanometer layer of conductive material, such as gold, from a sputtering machine; however, this process has the potential to disturb delicate samples.

Small, stable specimens such as carbon nanotubes, diatom frustules and small mineral crystals (asbestos fibres, for example) require no special treatment before being examined in the electron microscope. Samples of hydrated materials, including almost all biological specimens have to be prepared in various ways to stabilize them, reduce their thickness (ultrathin sectioning) and increase their electron optical contrast (staining).

There is a risk that these processes may result in *artifacts*, but these can usually be identified by comparing the results obtained by using radically different specimen preparation methods. It is generally believed by scientists working in the field that as results from various preparation techniques have been compared and that there is no reason that they should all produce similar artifacts, it is reasonable to believe that electron microscopy features correspond with those of living cells.

In addition, higher-resolution work has been directly compared to results from X-ray crystallography, providing independent confirmation of the validity of this technique. Since the 1980s, analysis of cryofixed, vitrified specimens has also become increasingly used by scientists, further confirming the validity of this technique.

Literally speaking, a microscope is a tool that lets the user see objects at a magnification greater than the actual specimen. The most common type of microscope is a magnifying glass, which uses a ground lens to focus the light reflecting off an object into a larger image. More complex light microscopes use a series of lenses to further magnify the object. However, conventional light microscopes have a resolution limit of approximately 250 nanometers (1 nanometer = $1X10^{-9}$ meter), approximately the wavelength of the incoming light used to illuminate the sample. This means that features on the sample smaller than this distance cannot be differentiated from each other. In order to see smaller details, a microscope that uses an illumination source with a smaller wavelength is needed.

High voltage electrons are the most commonly used source, although x-rays and neutrons can theoretically be used as well.

Fig. Picture of a Philips Transmission Electron Microscope with the Column and Viewing Screen Readily Visible.

There are two common types of electron microscopes: scanning (SEM) and transmission (TEM). When using an SEM, bulk biological samples are first coated with a metal that readily reflects electrons. This coating also provides a conducting surface for electrons to avoid charging of the sample. The incoming electron beam is condensed into a small beam which is scanned over the object. An image is formed by the electrons that bounce off the surface of the specimen and are then collected onto the imaging screen. The observer therefore sees a picture of the surface of the sample, without any internal information. A TEM, on the other hand, produces an image that is a projection of the entire object, including the surface and the internal structures. The incoming electron beam interacts with the sample as it passes through the entire thickness of the sample. Therefore, objects with different internal structures can be differentiated because they give different projections. However, the projection is of necessity two-dimensional against the viewscreen and relations in the z-axis between structures are lost. Furthermore, the samples need to be thin, or they will absorb too much of the electron beam.

LAYOUT OF A MICROSCOPE

In a sense, a transmission electron microscope works in much the same way as an optical microscope. At the top of the column, there is a high voltage electron emitter that generates a beam of electrons that travel down the column.

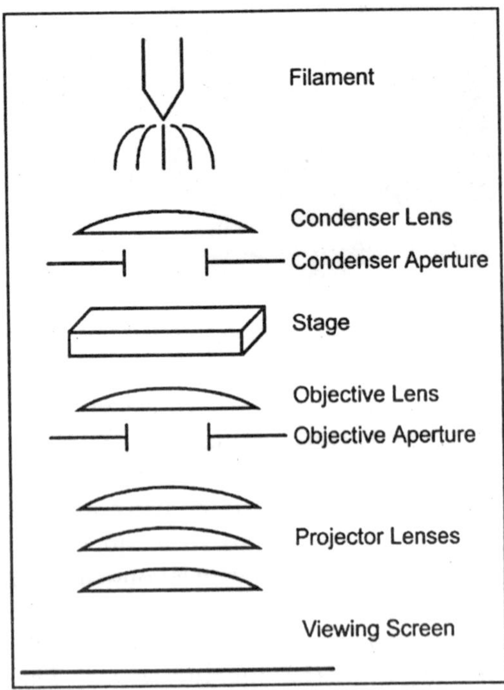

Fig. Cutaway Side View of an Electron Microscope Column, Showing the Principal Components.

These electrons pass through the sample and a series of magnifying magnetic lenses, to where they're ultimately focused at the viewing screen at the bottom of the column. Different lenses can be used to change the magnification and focal point of the image. Apertures along the column can be used to change the contrast and resolution of the image. The column itself is at a very high vacuum to minimize interactions between the electron beam and air molecules. Following are the principal parts of a TEM.

The Gun

Electrons are emitted from a wire that has been superheated by an electric current until enough energy (~2500K) is produced to overcome the work function of the metal, usually tungsten or lanthanum hexaboride (LaB_6). The emmitted electrons have a Boltzman energy distribution and need to be collated into a tight beam and sent down the microscope column, across the sample, lenses, and apertures. Thermionic and field emission guns (FEG) are the two most common electron guns used.

The Lenses

Electron lenses are magnetic coils that have been tuned to focus and direct a passing electron beam. Because the lenses are tuned for one particular electron wavelength, they will bend electrons of other wavelengths less evenly. Therefore, there is a splitting of the electron beam as it passes through the lens. This chromatic aberration effect is not unlike the splitting of sunlight into a rainbow by a prism. In the case of a prism, light of different wavelengths are refracted to different angles as they pass through the prism. Lenses can also rotate the electron beam if they are not tuned correctly. When this happens, the image can rotationally jump as different lesnes are inserted.

Three primary lenses are used to form and magnify an image. The objective lens is the topmost lens and does the first step of image magnification and focusing. It also works in conjunction with the objective aperture to generate amplitude contrast. The next lens is the intermediate lens and its positioning and strength controls the magnification of the image or the diffraction pattern. The last lens is the projector lens used to focus and project the image onto the imaging surface.

The Apertures

Apertures are holes along the microscope column that can limit the size of the electron beam that passes through it. Depending on it's location in the column an aperture can have different uses. The first aperture, the condenser aperture, for example, is located near the top of the column and as the name implies, is used to condense and mantain the coherence of the electron beam. The second aperture, the objective aperture, is located below the sample just after the objective lens. The objective aperture is used primarily to control contrast in the image. See below for a more thorough description of amplitude contrast.

The External Stuff

The outside of the microscope is just as important as the inner machinery. A typical user console consists of an informational screen and various control surfaces. These knobs, buttons, and switches can be used to control sample movement, beam brightness, beam current, alignment, focusing, and magnification, among other things.

The screen generally displays pertinent information on the functioning of the microscope, such as the current magnification, beam strength, defocus value, beam voltage, etc. Older microscopes use CRTs, but newer models are now interfacing with PCs to allow greater flexibility and control.

The sample grid is placed in a metal wand that is part of the goniometer or compustage. In figure, to the left of the compustage are the three control knobs for the three apertures. The tall white cylinder to the right and a bit behind the column is a cryo dewar that holds liquid nitrogen, used to cool down the column for easier removal of water vapour.

The two control pads, used to move the sample and beam are seen at the far bottom of the image. To the right edge of the image is the computer monitor that displays the microscope control software. In older microscopes, these controls were often mechanical knobs and buttons. Newer microscopes use a computer interface to directly access microscope functions.

When tilt microscopy is needed, it is the goniometer that rotates to change the angle of the sample. In the model shown in figure 1, the holder has been inserted into the goniometer. The large black casing on the holder is used for storing liquid nitrogen to keep the sample cold.

In normal use, there can be control wires attached to the goniometer/holder assembly that monitor the temperature for better performance.

THE ELECTRON GUN

The electron microscope generates images by bombarding a sample with electrons and then transforming the results of that interaction into a viewable image. For the image to form, the electrons that are generated need to be tightly controlled.

Types of Electron Guns

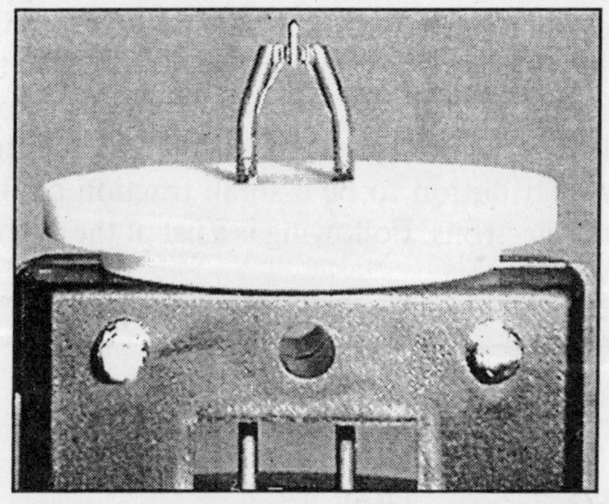

Fig. A LaB_6 Filament.

Heated Tungsten

A heated filament made from the metal tungsten. Much in the way that an incandescent lightbulb works, the high voltage that is fed through the filament causes electrons to be kicked off the filament. The amount of energy required is known as the work function.

Lanthanum Hexaboride (LaB_6)

The LaB_6 filament is also a thermal filament. However, its work function is lower than for a tungsten filament, so it is more efficient.

Tungsten Field Emission Gun (FEG)

The FEG gun is not a thermal filament. Instead, electrons are expelled by applying a very powerful electric field very close to the filament tip. The size and proximity of the electric field to the electron reservoir in the filament causes the electrons to tunnel out of the reservoir.

COHERENCE

Temporal Coherence

Because the emitted electrons in the various types of guns are heated, their energy distribution is not a sharp peak. Instead, they have a Boltzmann distribution that can vary widely depending on the type of filament. For a good microscope, you want “E/E to be as small as possible, that is, the energy distribution to be a small fraction of the average energy of the electrons. Following is a list of the average energy distributions:

Filament	Energy Distribution (“E)
Heated tungsten	2.5 eV
LaB_6	1.5 eV
Warm FEG	1.0 eV
Cold FEG	0.25 eV

It is clear that the FEG based microscopes have an advantage over the other types. For this reason, they are most often used for high resolution imaging. Ideally, cold FEGs are the best. However, they suffer from rapid degradation and are not economical.

Spatial Coherence

Spatial coherence basically means beam brightness. If the electrons are all emmitted as close to parallel as possible, then the beam will remain together over a long distance. This effect translates to greater brightness. The phase of the electrons also affects coherence. Ideally, the emitted electrons should all be in phase. If they are in phase, the electron waves interact constructively. If the phases are not aligned, then destructive interference will occur, and the beam will be less bright.

PRINCIPLES OF IMAGE FORMATION

When an electron beam strikes an object, several things can happen. If the electron does not strike an atom in the sample, it will continue to travel in a straight line until it hits the imaging screen. If the electron does come into contact with the sample, it can either bounce off elastically, that is, without any loss of energy, or inelastically, *i.e.*,

transferring some of that energy to the atom. In an inelastic bounce, the amount of energy transferred from electron to sample is variable and random. Therefore, when the electron eventually reaches the imaging plane, it has an unknown energy and angle of incidence. This particular electron will then generate noise in the image. However, if the electron bounced elastically, it's energy is constant, and the law of conservation of momentum will determine the angle at which it will bounce. This electron can be used to give high resolution information on the sample. Note that an electron can have an angle of reflection greater than 90°. Electrons that follow such a trajectory are termed 'back reflected electrons' and are generally not used in electron microscopy.

Contrast arises when there is interference between electrons coming in from different angles. Electrons that interact with the sample are bent away from their original path, and will thus interfere (either constructively or destructively) with the main electron beam. If a small objective aperture is used, electrons that get deflected at a greater angle are blocked, and the contrast of the image is enhanced. However, electrons with a high deflection contain high resolution information and are therfore lost. A balance needs to be achieved between having good contrast and having a high resolution.

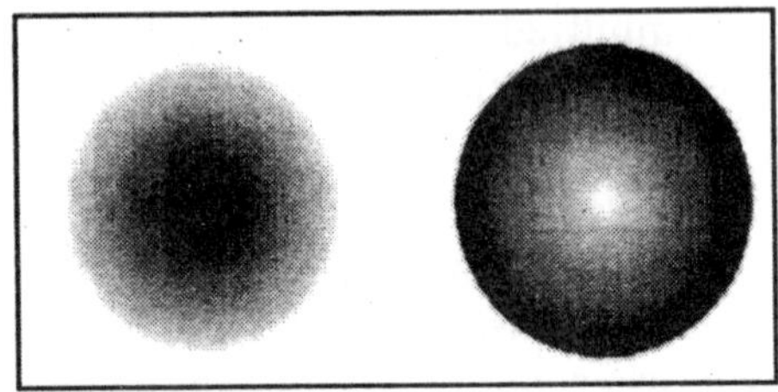

Fig. Differentiating between a Solid Ball (*left*) and a Hollow Sphere (*right*) in a Transmission Electron Microscope.

Because the electrons of a TEM pass through the sample before hitting the imaging screen, they contain information on the inside structures of that sample. If a hollow sphere and a solid ball are held in front of a flashlight, both will cast an identical shadow: a solid black disk. In a TEM, however, the two projections will be different. The solid ball will generate a disk that is very dark in the centre and progressively drops off in darkness towards the edges. The hollow sphere will also cast a circular image. However, the darkness gradient will be in the opposite direction, with the edge being the darkest, and the middle the lightest. This happens because the electron beam passes through the least amount of matter in the middle of a hollow sphere. The darkness of the image is proportional to the electron absorbency properties of the material used.

Electron microscopes can also be used to generate and view diffraction patterns of samples. These are usually from samples that contain a repeated motif, such as 2D crystals (sheets one layer thick) of proteins. In the back focal plane of the objective lens, a difraction image is formed. Depending then, on the positioning of the intermediate lens, the difraction pattern or the image will be magnified and displayed on the view screen.

GRIDS AND SAMPLE PREPARATION

Most biological EM work is done on small (several millimeters) copper discs called *grids* cast with a fine mesh.

This mesh can vary a lot depending on the intended application, but is usually about 15 squares per millimeter (400 squares per inch). On top of this grid, a thin layer of carbon is deposited by evaporating carbon graphite onto it. It is on this thin carbon film that the sample will then rest so that it can be examined in the microscope.

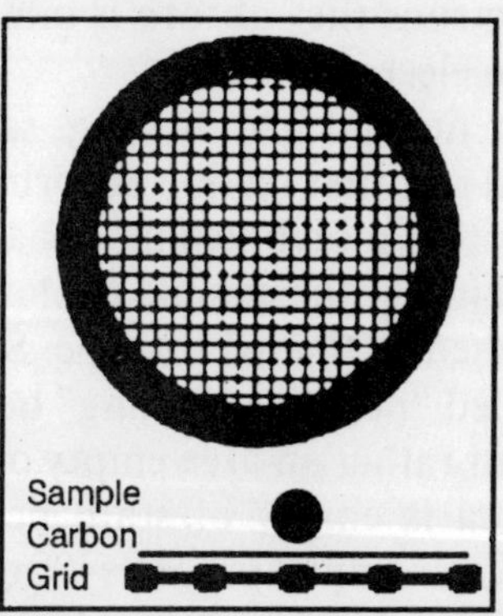

Fig. *Top:* Overhead View of a Grid, Showing the Mesh. *Bottom:* Side View of a Carbon Coated Grid Showing the Relative Position of the Carbon Film to the Grid and Sample.

Carbon is generally a hydrophobic substance (that is, it repels water), and if a drop of water is placed on it, the water will want to minimize it's contact with the carbon. To make the surface more accessible to water and the suspended sample, the carbon needs to be made hydrophilic.

This is accomplished by glow discharging. In glow discharging, the carbon coated grids are placed inside a partly evacuated chamber connected to a power supply. When high voltage is applied between the cathode and anode at each end of the chamber, the electron potential ionizes the gas within the chamber. These negatively charged ions then deposit on the carbon, giving the carbon film an overall hydrophilic (water attracting) surface.

After a small drop of the sample is placed on the hydrophilic grid, it needs to be stained so that the sample can be easily differentiated from the background. Transmission electron microscopy uses a high energy electron beam to bombard the sample. Depending on the amount of energy that was absorbed by the sample, the intensity of the beam that hits the viewing screen varies, and an image is made (remember that contrast arises from the beam interacting with the sample). However, carbon, oxygen, nitrogen, and hydrogen, the main components of biological molecules, are not very dense, and the amount of electrons they absorb is minimal compared to the intensity of the electron beam.

Therefore, for normal EM viewing, samples are stained with a heavy metal salt that readily absorbs electrons. This is usually lead, tungsten, molybdenum, vanadium, or depleted uranium. After staining, the sample is blotted, air dried and ready to be examined in the microscope. Staining with these heavy atoms is called "negative staining" because one sees not the object itself, but rather an area empty of stain surrounded by stain. This area is empty of stain because the sample material (protein for example) prevents it from depositing onto the carbon layer. Good contrast is achieved when the stain completely surrounds the specimen and neighboring area.

Holey Grids

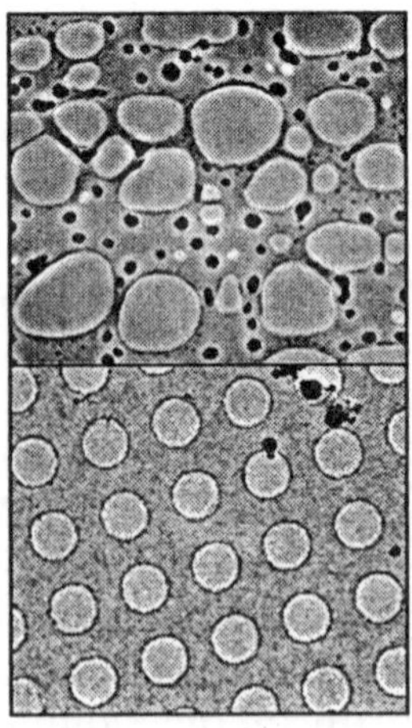

Fig. Top: Overheard Image of a Homemade Holey Grid. *Bottom:* Image of a Manufactured Holey Grid.

Although the carbon film is very thin and usually presents

no big problem in obscuring a sample, for high resolution studies of unstained biological macromolecules any interference on the beam can be problematic. For high resolution studies, holey grids are used in conjunction with freeze-sample (cryo) electron microscopy. Similar to normal carbon coated grids, holey grids are covered with a fine layer of carbon.

However, as part of the preparation process, the carbon film is deposited in such a way that there are holes of a desired size in the carbon (figure 2). One of the purposes of these holes is to eliminate any absorption and scattering of the electron beam by the carbon film, which will generate noise and obstruct the signal. Because cryo EM does not use staining, any elimination of background noise is desireable. The holes also allow for "pockets" of solvent to form. Within these pockets, the specimen remains fully hydrated, even when the sample has been frozen.

A problem can arise when holey grids are used in conjunction with proteins that are positively charged, such as DNA binding proteins. Because carbon film has a slight negative charge, especially after glow discharging, it readily attracts positively charged proteins. Therefore, such a protein will preferentially land on the carbon as opposed to the holes. A workaround has been developed using a secondary carbon layer. A holey grid is made as described above. Before depositing the sample, however, a very thin layer of carbon is deposited above the original film. Because this carbon is continuous, there is no preference to where the protein will lay down. Furthermore, the layer is thin enough to not have a great impact on the signal to noise ratio of the sample.

CARBON EVAPORATION

Samples for electron microscopy are placed on a metal support grid which needs to be first covered with a low absorption material. This is usually a thin film of carbon which has been evaporated onto the metal grid from a graphite rod. The specimen is then deposited on top of this carbon, stained, and blotted dry.

The principle of carbon evaporation is very straightforward. A sheet of freshly cleaved mica to be carbon coated is placed within an evacuated chamber that has a small graphite rod connected to a high voltage circuit. A well-cleaved mica surface should have an atomically flat surface, perfect for high resolution work. When the chamber with the mica sheet (or sheets) has been evacuated to a sufficiently high vacuum, current is applied across the graphite rod until it incandesces. At this point, carbon is evaporated off the rod and deposited on the surface. After sufficient carbon is deposited, the current is turned off and the coated mica retrieved.

Grids to be carbon coated are submerged onto filter paper in a small water tank. The coated mica is then slowly and carefully lowered into the water, in such a way that the carbon film slides off and floats at the surface. When the water is removed from the tank, the carbon follows the receding water level to cover the grids. The grids are now ready for use after drying.

NEGATIVE STAINING

For many types of microscopy, the contrast of the sample is too poor for the human eye to easily differentiate edges and features of the sample. In order to improve contrast, several techniques are used. For light microscopy, phase contrast microscopy is an often used technique. In this method, the incoming light suffers a phase shift as it interacts with the sample.

Because the shifted wave interferes with the rest of the beam, contrast is created, and edges stand out. Another common contrast enhancement method uses stains to preferentially highlight some parts of the sample over others. Common stains include colored dyes and highly absorbent materials. For stains to be effective, they have to preferentially bind to some parts of the sample, thereby allowing the differentiation of different parts.

In electron microscopy, staining is usually done with heavy metal salts commonly derived from molybdenum, uranium, or tungsten. Heavy ions are used since they will readily interact with the electron beam and produce phase contrast.

A small drop of the sample is deposited on the carbon coated grid, allowed to settle for approximately one minute, blotted dry if necessary, and then covered with a small drop of the stain (for example 2% uranyl acetate). After a few seconds, this drop is also blotted dry, and the sample is ready for viewing.

Contrast from Interaction of Electron Beam with Stain

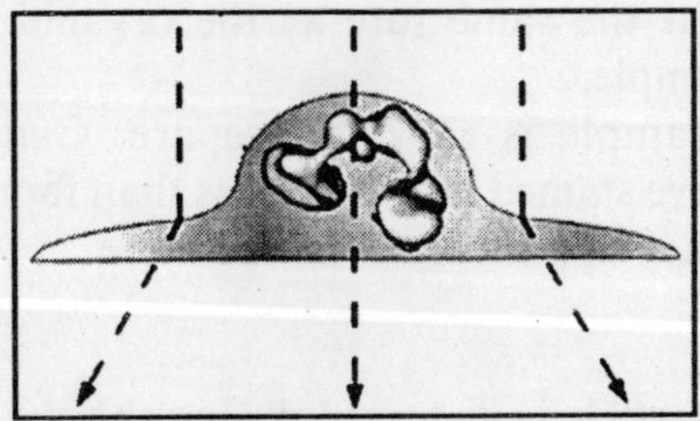

Fig. A Sample Deposited on a Carbon Coated Grid and Surrounded by Stain (Light Gray Shading) Interacting with an Electron Beam (Arrows).

As explained earlier, the interaction of the electron beam with the sample generates contrast in the image. In negative stain microscopy, the electron beam primarily interacts with the stain. When stain is added to a sample, the stain surrounds the sample but is excluded from the volume occupied by the sample; hence the use of the term 'negative stain'. As figure 1 shows, a well-stained sample is uniformly covered by the stain.

When the electron beam (arrows) passes through the sample, it will be deflected by its interactions with the sample and stain. Since the protein sample excludes stain, the deflection of the electron beam through protein (centre arrow) is less than that through stain rich regions (outer arrows). Electrons that are highly deflected by the stain are then filtered out by the objective aperture located below the sample. Depending on the size of the aperture, the quantity of electrons that are culled out will vary and determine the contrast and resolution of the image.

Drawbacks and Benefits of Negative Staining

Benefits

- Very high contrast. The stain absorbs electrons in much higher amounts than the surrounding medium. Therefore, different regions of the sample have different electron densities and can be differentiated easier in the resulting projections.
- The radiation damage from the absorbed electrons is mostly irrelevant because it does not affect stain salts at the same rate as the organic molecules of the sample.
- The sample is easy to prepare. Generation of a negative stained grid takes less than four minutes and no fancy apparatus.

Drawbacks

- The particle is distorted during the staining process. As part of the drying processes, the particle loses it's hydration shell. Often, this shell stabilizes the soluble particle onto a certain configuration and deposition on the carbon can cause it to change shape.
- Artefacts can arise if the stain is uneven. The pattern of stain deposition is dependent on the structure of the particle. Certain areas might acquire more stain and therefore appear with higher contrast than would be normal.
- The resolution is limited to approximately 20 E under optimal conditions. This corresponds to the size of the salt grain used for staining.

CRYO-FREEZING THE SAMPLE

Cryo EM is a microscopy technique in which the sample to be viewed is frozen in a very cold liquid refrigerant in order to preserve and protect it during observation. Biological molecules need a solvent to be stable. In most cases, a water/ salt solution is enough. Evaporation, however, needs to be eliminated when a sample is inserted into the electron beam. This can be accomplished by either blotting the sample dry for negative staining, or by freezing.

Frozen Samples

Freezing a sample is as simple as dropping a prepared grid into a very cold storage medium. The solvent water around the sample is frozen in place, and the sample is cryogenically protected. In actuality, however, the process is not this simple. First, the freezing process must be quick enough to prevent the frozen water from forming cubic ice. Cubic ice is water in a crystal lattice that readily absorbs the electron beam, obscuring the sample.

Preventing cubic ice can be accomplished by using a sufficiently cold freezing agent. If the sample freezing is quick enough, the water will only solidify as an amorphous solid (vitreous ice) and not have a chance to crystallize. Liquid nitrogen is at a cold enough temperature (approx. -195 °C) to speed the process. Unfortunately, it's heat capacity is very low. Therefore, as soon as a grid at room temperature is dropped into the liquid nitrogen, some of it will warm and boil off, slowing the freezing process and allowing cubic ice to form. Ethane has a much higher heat capacity than liquid nitrogen and is also liquid at temperatures just slightly above those of liquid nitrogen. Therefore, liquid ethane is cold enough (it's melting point is -188 °C) to freeze water quickly and correctly, while not boiling off in the process.

The very properties that make liquid ethane such a favorable cryogen also make it dangerous to work with. Some safety issues must be dealt with before using.

Sample Preparation

As mentioned above, a biological sample is quickly frozen to preserve it's hydrated state. The following illustrates one example of how the sample is immersed into the cryogen for freezing.

A small vial of ethane (cryogen tank in figure 1) is placed inside a larger liquid nitrogen reservoir. An EM grid is held in place at the bottom of a plunger (usually by fine tweezers). The plunger has a heavy weight at the top for extra force. Once the ethane in the vial is completely frozen, it needs to be slightly melted.

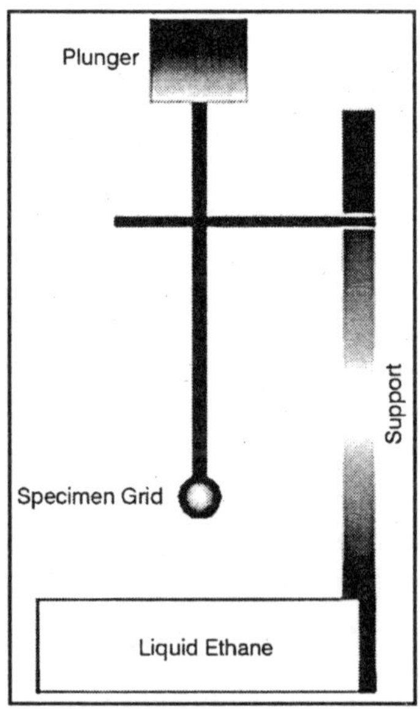

Fig. Schematic of a Freeze Plunger. A Drop of the Sample is Placed at the End of the Plunger and Rapidly Immersed into the Cryogen Tank.

When this small volume of liquid ethane is ready, a few microliters of the sample is placed on the grid. A piece of filter paper is then pressed against the sample to blot of the excess buffer.

After a predetermined time, the filter paper is removed, and the plunger is allowed to drop into the liquid ethane. Once the grid enters the liquid ethane, the sample is rapidly frozen, and the grid can then be moved to a storage box in liquid nitrogen for later use in the microscope.

Several modifications are common to this procedure. First, the blotting and plunging can be done in a high humidity temperature controlled chamber. This allows the operator to more carefully control the amount of solvent that is left on the grid. If too much water remains, the resulting ice will be too thick for observation. If too much water is removed, the sample will interact with the surface and perhaps be disturbed. Therefore, having the right amount of solvent is of utmost importance.

Fig. The Vitrobot™, Produced by Maastricht Instruments, is Computer Controlled to Simplify and Automate the Vitrification of Liquid Samples.

Another common modification involves changing the biological conditions just before plunging. Since some proteins can react to changes in their environment with millisecond resolution, manually changing the conditions before plunging will not work.

In these cases, some research groups have designed spray mechanisms that spray the plunging grid with a different buffer just before it enters the cryogen.

For example, the acetylcholine receptor has been imaged with and without an acetylcholine ligand using this mechanism.

The release of the ligand is too quick, so acetylcholine is sprayed on the grid just before it is frozen.

Fig. The Freezing Chamber of Vitrobot™ with a Grid Suspended before Freezing, and after Dunking into Liquid Ethane. The Outer Ring Contains Liquid Nitrogen to Keep the Entire Chamber at the Right Temperature.

CRYO ELECTRON MICROSCOPY STUDIES

In the preceding section, we covered the why's and how's of freezing a sample. This section explains the actual process of imaging a frozen sample. Also covered are other considerations that must be kept in mind when deciding whether to use cryo EM.

Imaging The Sample and Taking Pictures

While taking a picture of a negative stained sample is straighforward, imaging cryo EM samples is more involved. The sample is so delicate that the only time it can be exposed to the electron beam is when the picture is to be taken. Therefore, all image focusing and alignment must be done off to the side, in an area that is not going to be used. For more information on this low dose technique.

Low Dose

An electron beam is a stream of high energy particles that are bombarding the sample. The image that is viewed is a result of the interaction of the sample with this beam. Most of the electrons that form the high resolution image appear due to elastic scattering, where only their trajectory has been changed, but their energy is unaffected.

However, a small fraction of the electrons transfer some of their energy to the sample. This energy accumulates and can break apart molecular bonds, destroying the sample after some time.

Therefore, for high-resolution imaging, low dose parameters require that the area to be imaged is not exposed until the picture is actually taken. All image calibration and focusing is done beforehand on a nearby area, in the hope that it's properties are similar to the final imaged area. Also, for the final imaging, very low electron doses on the order of 6-10 electrons per E^2 are used. As a comparison, high-resolution electron microsopy of semiconductors routinely uses doses of 100 *thousand* electrons per E^2.

As a corollary of using a low electron dose during imaging, the signal to noise ratio of the image is very low. For image reconstruction, a proportionally larger number of molecules must therefore be imaged.

Advantages & Disadvantages Over Negative Staining

What are the advantages of cryo EM over negative staining?

- The sample is always in solution and never comes into contact with an adhering surface. Therefore, the shape that is observed is the true shape of the hydrated molecule in solution and has not been distorted by attaching itself and flattening against the supporting film.
- There is no stain to distort the sample. Stain does not always lay down evenly, which can generate artifacts and false contrasts when reconstructing the structure of a sample. Also, the staining process requires that the sample be blotted dry. During the drying, the sample can be damaged in many ways, such as by flattening and twisting.

- When the sample adheres to the carbon grid, it could stick in a preferential orientation. If this happens, then information will be missing from the final image set (a missing cone), and the resolution of the calculated model in that direction will be lacking.
- Low dose methods are normally used. Therefore, the electron beam causes less damage to the sample.

Are there reasons, then, to not use cryo EM?

- Very low signal to noise ratio. Biological macromolecules are normally made up of carbon, hydrogen, oxygen, and nitrogen. The electron absorption of such molecules is very low. As a result, image contrast is also very low and it is hard to detect features when dealing with just a few images.
- Can not do tilt imaging as well. The ice cross section of a tilted frozen sample is too thick to yield good images. Only untilted images are usually taken. Furthermore, charging is more widespread when imaging a tilted frozen sample.
- More time consuming to generate samples. However, this is generally not a big problem, especially once a working protocol is designed and good samples are readily available.
- If vitreous ice cannot be easily formed, the resulting cubic ice absorbs electrons very easily and the frozen sample is basically worthless.

Other Considerations

One of the reasons for using cryo techniques in electron microscopy is that the specimen is frozen in time. Thermal vibrations are reduced to a minimum, and the corresponding noise is removed from the images. When a specimen is bombarded with electrons, some of these electrons are absorbed into the surrounding medium (vitreous ice in the case of a good cryo sample). There they energize the ice (also known as charging) and induce vibrations, which can move the embedded sample. This movement reduces the resolution that can be achieved and also damages the sample. To avoid this problem, it is useful to capture part of the holey carbon grid within the electron beam. At high enough voltages, carbon can conduct charges over its surface. When the electron beam charges the ice, the carbon can act as a ground sink, bleeding surface electrons away from the delicate sample.

BASICS OF SINGLE PARTICLE RECONSTRUCTION

In short, the general purpose of single particle reconstruction is to generate a three-dimensional model from many two-dimensional images. This, of course, is a general goal, as a lot of information can be gathered from the various intermediate steps. Regardless, for the purpose of this tutorial, we will follow the steps necessary to generate a 3D model.

Methodology

In order to most clearly understand the various steps that will be discussed, it is important to know what the goal is and what our starting conditions are. The data typically consists of micrographs containing projections of proteins or protein complexes generated in a transmission electron microscope. From these experimental projections (particles), we will generate a 3D model. Later on we will discuss how the actual model is generated, but for now we just need to know that the process requires the assignment of Euler angles to each experimental projection. Once these angles are assigned to each experimental projection, the model can be generated.

Euler angles are directional angles used to indicate a position and orientation in space around a common centre. Their definitions are as follows:

θ (theta)

Defines the elevation above or below the equator.

φ (phi)

Defines the rotation (azimuth) around the equator.

ψ (psi)

Defines the rotation around the centre of the position defined by θ and φ.

If one imagines standing in the centre of a large hollow

sphere, in order to define a point on the surface of the sphere, one needs to know two pieces of information: what the elevation of the spot is relative to the equator of the sphere (θ) and what the rotation is around the vertical axis of the sphere (φ - azimuth).

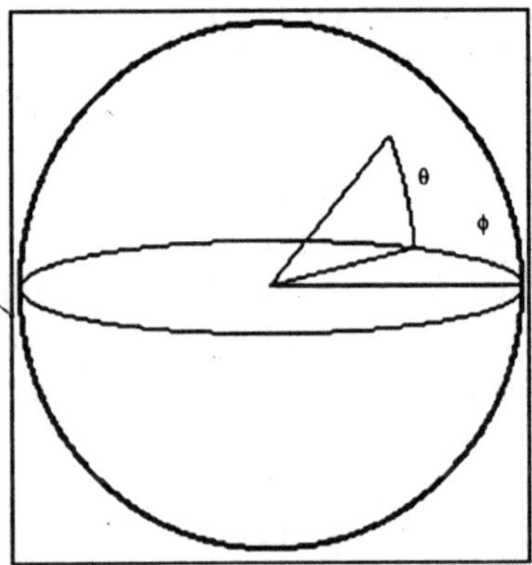

Fig. Diagram of an Euler Sphere Indicating How θ and φ are used to Describe Any Point on the Surface of the Sphere.

To generate a reconstruction, we will need to determine θ and φ for each experimental particle in order define its position on the Euler sphere. The centre of the Euler sphere represents where our 3D model is. Assignment of the Euler angles is most commonly done by the method known as random conical tilt.

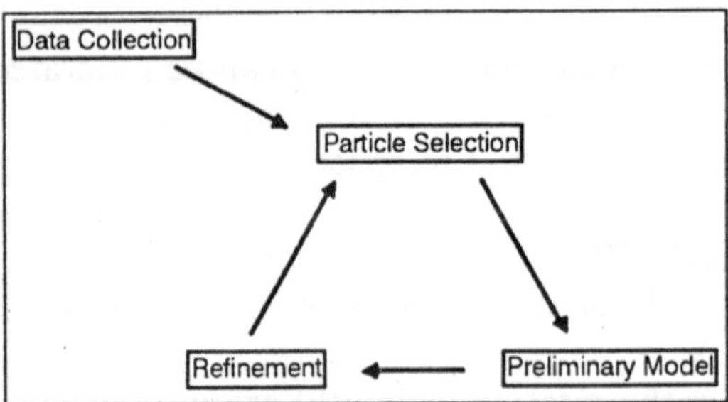

Fig. Flow Chart of a Typical Reconstruction. The Process is Designed to be Iterative.

The process itself is very iterative. Experimental projections are first selected from electron micrographs. These particles are then centered, aligned, and classified. From the centered/aligned/classified particles, a preliminary model is generated. To refine this solution, the model is used to better align and classify the original particles. A new model is generated from the refined data, and so on until a satisfactory solution is reached.

PICKING PARTICLES

Picking Particles from Tilt Pairs

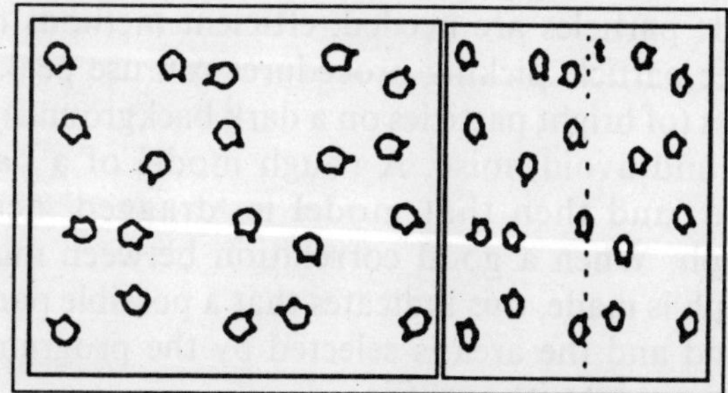

Fig. An Untilted (*left*) and Tilted Micrograph (*right*) Pair.

In order to reconstruct a 3D molecule from it's 2D projections, it is often necessary to use several thousand experimental projections (particles). For small sample sizes (e.g. less than two thousand), it is straightforward to select (pick) the particles from several micrographs. Using a computer programme such as SPIDER/WEB, the untilted and tilted images (for random conical tilt) of the same micrograph are displayed side by side. Several particle pairs are then picked by hand. From their geometrical relationships, this allows the programme to determine the angular relationships between the pairs (the exact tilt between micrographs and any rotation of the micrographs perpendicular to the tilt).

After the angles have been determined, usually only the untilted particle need to be selected by the user, as the location of the tilted particle is calculated automatically using the previously calculated angles. Of course, it is wise from time to time to recalculate the angles, however, as drift can occur from one area to another of the micrograph. This procedure is repeated for as many micrograph pairs as there are. In reconstructions that do not need tilt pairs (as when a reference is used, or when using the common lines method), picking particles is much easier, since any suitable particle can be used (there is no need for a matching tilted pair).

Automated Particle Picking

This procedure, of course, can become tedious after several dozen micrographs have been scanned. There is no real easy way to automatically pick all particles in tilt pairs, but when only single particles are needed, efficient methods do exist. Automatic particle picking procedures can use peak finding algorithms (of bright particles on a dark background) to select particles and avoid noise. A rough model of a particle is generated, and then that model is 'dragged' across the micrograph. When a good correlation between model and micrograph is made, this indicates that a possible particle has been found and the area is selected by the programme and windowed out into its own file.

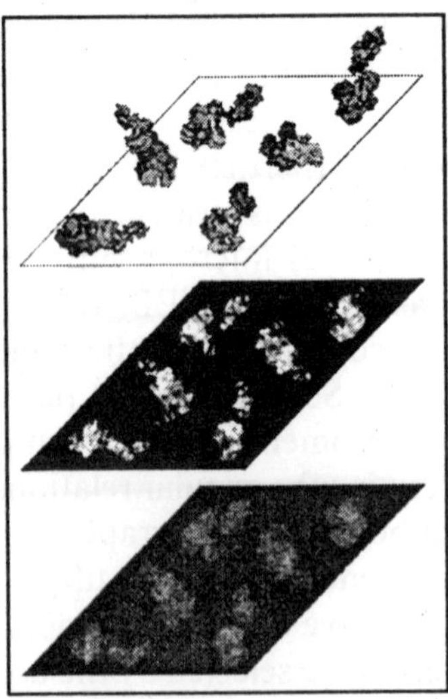

Fig. Top: Particles of DNA Polymerase Randomly Strewn on a Surface. *Middle*: Computed Two Dimensional Projections of the Above Particles. *Bottom*: The Same Projections with Noise Sdded to Indicate the Loss of Detail that is Seen in a Microscope.

Another common method of automatic picking uses two steps: training and selection. In the training step, a small amount of particles are picked by the computer, and then the operator manually separates them into particles, junk, and background noise. The computer uses this information to design a discriminant function and then proceeds to pick the particles. With enough training, good programs should be able to differentiate between the various possibilities.

Sometimes, too, the procedures can be calibrated to pick particles that are more than a set distance apart in order to avoid overlaps and picking the same particle twice. Unfortunately, these procedures generally only work on sample preparations that are homogenous. Large particle aggregates, as well as other artefacts in the scanned image, can confuse the picking. For these reasons, there is often a quality control step where a person manually scans the picked particles to ensure a low level of false positives. Nevertheless, once all the particle coordinates have been determined for the micrographs, the particles are 'windowed' out into their own individual files, where they can be manipulated in a more efficient manner. Some programs simplify the task of keeping track of so many particles by 'stacking' them into one file. In a stack, all the images are concatenated into a continuous data stream, where it is easier (and quicker) for the programme to access.

ALIGNING PARTICLES

Centering the Particles

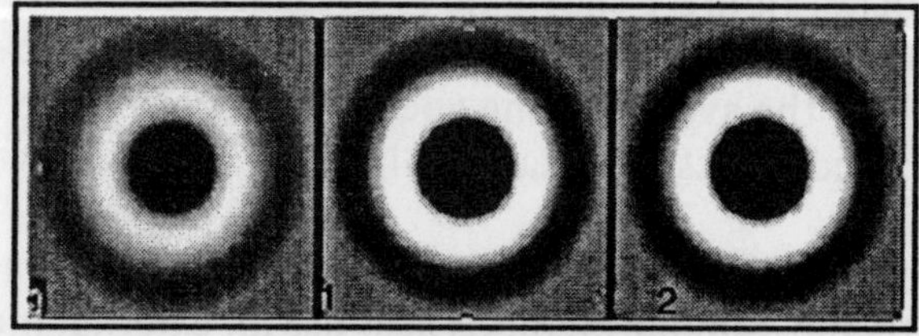

Fig. Averages Showing the Stepwise Centering of Several Ring Particles. Notice how the ring becomes less diffuse with each iteration, indicating better centering of all the particles. In each image, the particle is white on a dark background.

Usually, most picking algorithms (whether automated or manual) do a fairly good job of placing the picked particles near the middle of the boxed out image. However, fine adjustments still need to be done to make the later steps all that more efficient. A common method of centering searches for the so-called centre of gravity, In this method, the image is scanned and the intensity peaks calculated to find the centre of the particle 'mass'. The image is then shifted by the correct amounts to maximize the number of peaks near the centre.

This method, however, does not work very well if the particle is very asymmetric, since one end might be very heavy compared to the other. In this case, the heavy end will be centered, and the rest of the particle might not necessarily align correctly. Also, particles that have a hollow, such as in a ring, can be miscentered, since some centre of gravity algorithms break down when there is a lack of mass in the middle of the object. In these cases, the ring particles are kicked off centre so that part of the mass of the ring is in the image centre.

Another centering method averages all the picked particles together, and then crosscorrelates each individual particle to the average. The crosscorrelation function is used to determine by how much to shift each particle when trying to centre it. When all the particles have been crosscorrelated to the average and shifted, a new average is generated. Once again, all the particles are compared to the new average, and shifted as necessary to centre them as best as possible. This iterative process is repeated until no significant shift is necessary for all the particles. figure 1 shows three averages produced during the iterative process of centering 590 ring-shaped particles. One can see how the averaged ring goes from very fuzzy to having well defined edges, indicating that the data set is nearly centered.

Crosscorrelating to a global average is but one variation on this theme. Similar methods also use an external model or a rotational average of the particle itself as the centering reference. Unfortunately, it can be difficult to obtain a reasonable external refernce, so a global average or a rotational average are most often used.

Aligning the Particles

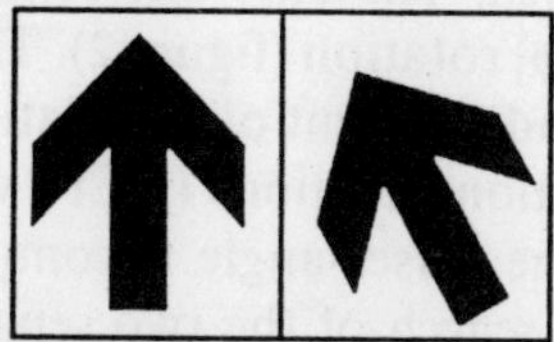

Fig. An Unrotated (*left*) and Rotated Plus Shifted (*right*) Particle Pair.

Once particles have been picked and centered, they need to be aligned, ordered, and classified for the reconstruction. These steps provide information on the relationships between different particles in a data set. The information can then be used towards putting together a three-dimensional model. For example, in the random conical tilt reconstruction method, the actual reconstruction is done from the tilted particles, but the alignment and classification comes from the untilted images. By determining the angles between untilted and tilted particles, it is a straighforward transformation to go from one coordinate plane to the other in later steps.

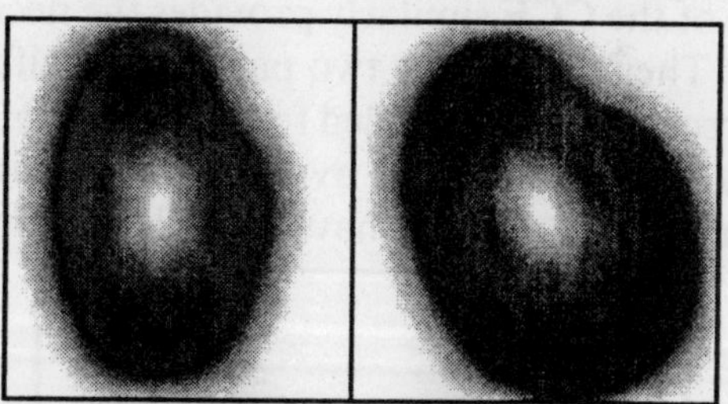

Fig. ACFs of an Unrotated (*left*) and Rotated Plus Shifted (*right*) Particle Pair.

Alignment involves placing the images of particles into a similar orientation on the screen. In the simplest case, two particles are just shifted (by [*x*, *y*]) and rotated (by [φ]) relative to each other in the plane of the image. Correlation, a method for calculating similarities, is used to determine the [x,y,φ] parameters. The correlation can be done either between images in the data set (crosscorrelation: CC) or by comparing an image to itself (autocorrelation: AC). Usually a mixture of the two is used, as described below. For this example, two similar objects will be aligned. The only difference between them is an [x,y] shift and a [φ] rotation (figure 2). The autocorrelation function (ACF) is independent of translation, so in this case, the two autocorrelation functions (ACF) will just be rotated by the angle [φ]. The offset angle is computed by doing an angular correlation search of the two separate ACFs. When this angle is calculated, one of the two original particles is then rotated by the offset angle, in order to align it with the other particle.

Fig. An Unshifted (*left*) and Shifted (*right*) Particle Pair.

With the two particles now in the same orientation, they need to be shifted in-plane to overlap. Crosscorrelation (CC; like autocorrelation but between two images) coupled with a peak search of the CCF similarly provides the necessary [*x*, *y*] shift values. Then, one of the two particles is shifted by [*x*, *y*] to get a correct overlap. If needed for future study, the aligned images can be summed and averaged. This procedure is repeated for all particles that have a similar orientation.

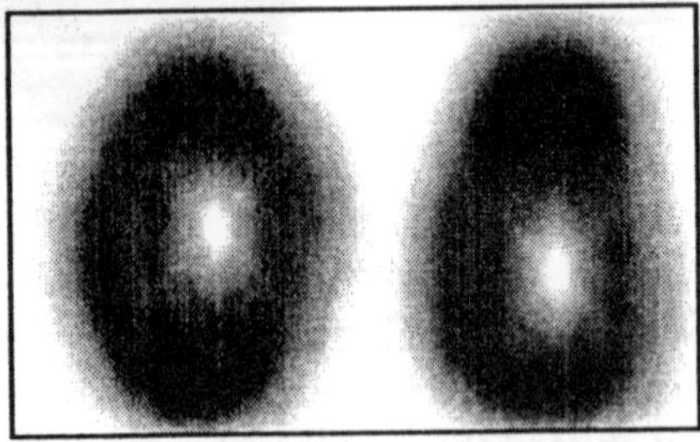

Fig. ACF (*left*) and CCF (*right*) of a Shifted, Unrotated Particle.

CLASSIFYING PARTICLES

Particle classification involves grouping images that are similar, and separating images that are distinct. In practical use, this means that experimental projections that have the same orientation (shape) are placed within the same category for later averaging. In this case, orientation means that the particles are showing the same face to the viewer and the only difference between them is that they can be rotated by some angle in the plane of the image. The experimental projections might also be shifted relative to each other, but the centering of the experimental projections is often done before classification.

Since protein particles generally do not lie in one preferred orientation on the carbon film, many views of the protein will generally be visible. Each of these classes of experimental projections has to be defined. Furthermore, alignment of particles then has to be done for each class.

For example, a book dropped randomly on the floor will preferentially land on its front or back, and then with less probability on one of the narrow sides, and finally with even less probability on just an edge or corner. The two principal classes (front showing and back showing) will make up the bulk of the data, and, as will be shown later, by having two orientations to work with, will help create a better reconstruction than if only one view was available.

Within a particular view, it is important that as many projections are used that cover the 360° rotation on the plane of the micrograph. This is particularly important when using the random conical tilt method. When the tilted projections are then used to generate a first volume, the entire space of information is available. If too few particles are used or there is a preference for one orientation over another within one class, the missing information will reduce the resolution of the calculated volume.

CREATING A MODEL

In the previous sections, we discussed how experimental particles are picked, aligned, and classified. In this section, we will briefly explain the two methods most commonly used for generating a three-dimensional model via single particle reconstruction.

Random Conical Tilt

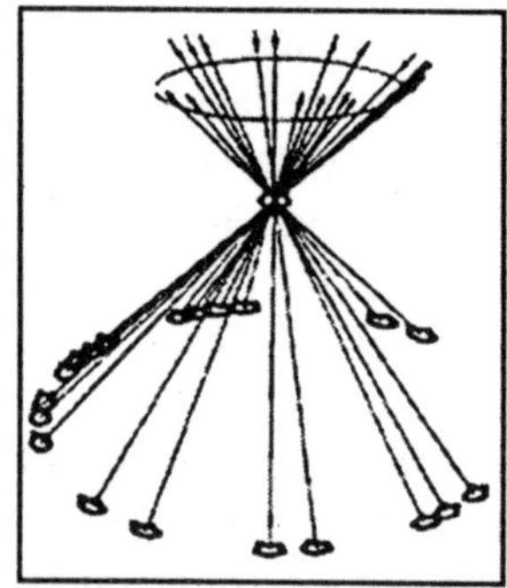

Fig. Principle of Random Conical Tilting, Showing How Many Rotated Images Within a Cone Come Together to Form a Surface.

For a specific class of particles, once the untilted particles are aligned with each other, those angles and shifts are applied to the tilted pairs to determine their relative orientations in space. As seen in figure 1, the projections of these aligned particles will form a cone with a fixed angle (the tilt angle) and a random azimuths (rotation) around the cone, matching the in-plane angles found in the alignment of the untilted particles.

Common Lines

Another reconstruction method searches for the intersection of any two projections in Fourier space. The Fourier transform of the experimental projections all form slices around a common core in Fourier space. Therefore, the intersection of these projections are unique (unless the projections perfectly overlap), and their relative orientation can be found when three or more projections are used. A principal problem with this method is that the handedness of the image is lost. This, however, can later be corrected by visual examination of the model with other known structural information.

Other Methods

There are several other methods that can be used for

generating three-dimensional models from data acquired by EM. They include:

Helical Reconstruction

Helical reconstruction is used when the protein of interest forms a natural helix. Since the helix is a recurring structure with a very well defined pattern, the repeating pattern of the helix can be exploited to solve the structure. In this case, no alignment of the particles is needed, since the individual positions of subunits within the helix are clearly defined by the shape of the helix. Two common examples of structures solved by helical reconstruction are TMV and microtubules.

Icosahedral Reconstruction

Icosahedral reconstructions also take advantage of internal symmetry and repetition to generate a detailed three-dimensional structure from the data set. In this case, the symmetry is icosahedral (twenty-one sided). Many viruses exhibit icosahedral symmetry in their capsid proteins, and this method has been used to solve their structures.

Electron Crystallography

Electron crystallography is similar to x-ray crystallography in that it exploits the repeating pattern found within a crystal to generate a structure. Just as with x-ray crystallography, difraction patterns are generated and are used to define an electron density map. However, it differs in that the crystal used is a two-dimensional sheet as opposed to three three-dimensional crystals of x-ray crystallography.

However, these methods do not generally deal with single particles, but rather with repeating arrays of particles. Therefore, no further mention will be made of them here.

BACK PROJECTION

Back projection is the process by which we generate our 3D model. As was discussed earlier, one of the goals of single particle analysis is to use 2D images formed by TEM to reconstruct a 3D model of the original object. In this section, we will see how the generated projections are "added up" to recreate the model.

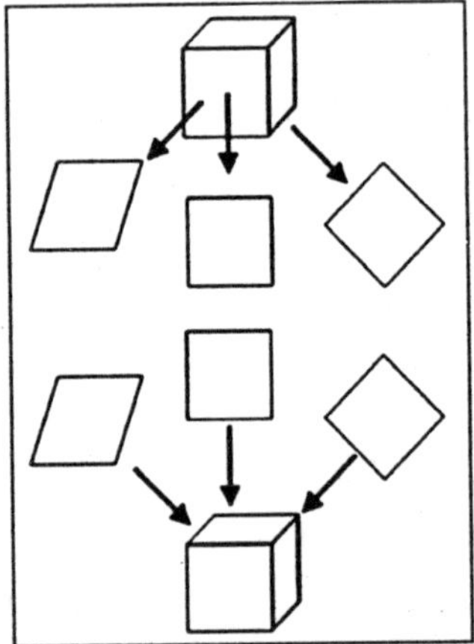

Fig. Example Image of How a Cube Generates Different Projections Depending on The Angle of Projection (*top*). These Same Projections can then be Used to Regenerate the Cube Through Back Projection (*bottom*).

As its name implies, back projection is the inverse function of projection. When an n-dimensional object is projected, each projection is an n-1 dimensional sum of its density along the projection axis.

Therefore, a sphere would have circles as its projections. A cube, on the other hand, would produce either squares, diamonds, or other intermediate parallelograms depending on the direction of projection. The actual shape, of course, depends on the orientation from which the projection was made. The reverse function is called back projection and regenerates the original object.

Since a projection can be thought of as a "squishing" of the object, back projection is then the "smearing" of the projections back onto each other.

The sum of the back projected projections regenerates the original object. In the example above, a cube generated different paralleograms in projection. In figure, back projected circles were used to reconstruct the original sphere. When a circle is "smeared" along its axis towards the centre of the space, it forms a line. In each frame of the animation in figure 2, more of these back projected lines are added at different angles.

As the amount of projections used is increased, the lines complement each other where they intersect. The result is that the sphere at the intersect point becomes better defined. Thus, the back projection of the projections can reconstruct the original object.

Fig. Animation showing how back projected 1D projections (white lines) can be merged to form a 2D object (white circle in centre). As more projections are added, the original object becomes more defined.

Back Projection with a Detailed Example

The following example presents a more detailed view of back projection using a more complicated model with extra features and details.

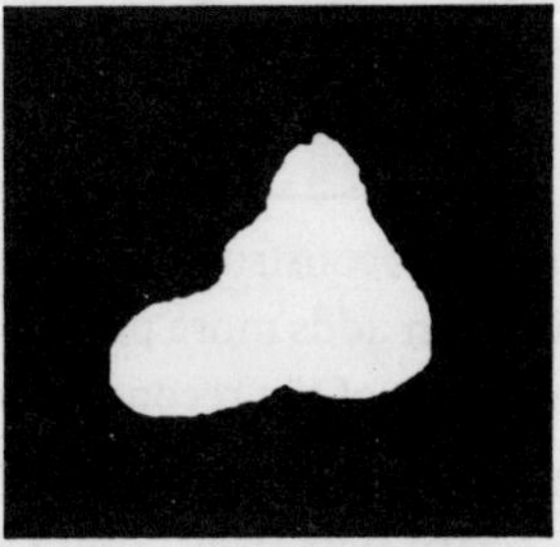

Fig. Picture of The 2D Model that Will be Reconstructed by Back Projection.

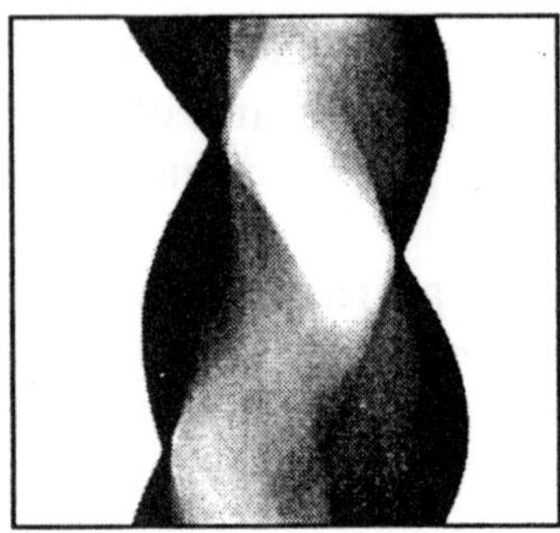

Fig. Sinogram Generated by Applying a Radon Transform to the Model. Each Row of the Sinogram Represents a Diffrent Projection Through the Model.

The Radon transform is an algorithm that transforms an original image into a series of equiangular projections. When applied to a 2D object, the output of the Radon transform is a series of 1D lines. For easier representation, the lines are often stacked on top of each other, generating a sinogram. The sinogram in figure 4 is the Radon transform of figure 3.

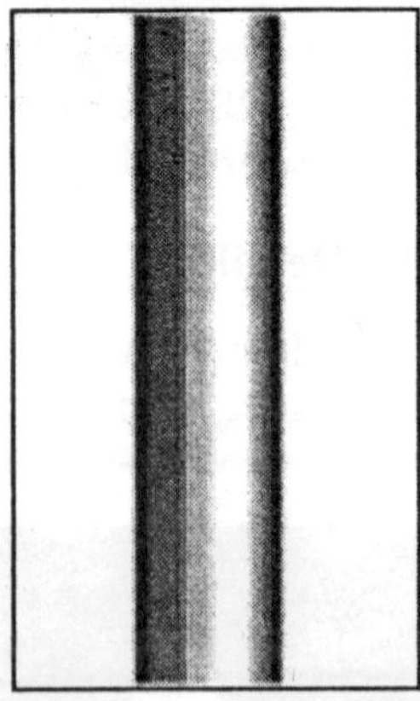

Fig. Animation of the 2D reconstruction of the model from figure. Each frame of the animation adds more projections that help improve the quality of the reconstruction.

To create the different figures for this example, the following SPIDER script was used: b00.radon.spi. This script generates a Radon transform from the input model, extracts each individual projection from the sinogram, and modifies the projections so that they can be averaged. While the script is not a mathematically accurate description of back projection, conceptually it is similar. Finally, it can be easily modified for other presentational examples.

Back Projection in the Real World

The above descriptions dealt with an ideal sample. All the information from the original model was known. This made it easy for the reconstruction programs to recreate the original model with high fidelity. But what happens with a real world example? Data is often missing or inaccessible.

In the previous steps of the reconstruction process, the experimental particles were centered and aligned. Lastly, each experimental projection was assigned a set of Euler angles that define its position in space relative to its origin from the original model.

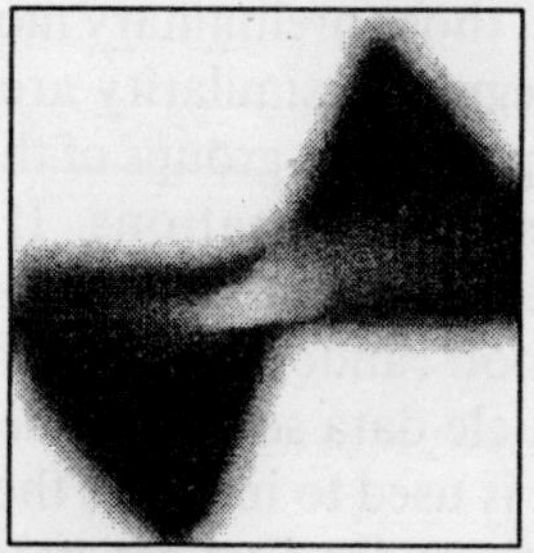

Fig. The Reconstructed Model with Only Half of the Projections Used. Since the Projections only Spanned Part of the Available Angles, Some Features are Sharper in One direction than in the other. Smears are visible in the directions that were overrepresented.

To generate figure 5 all the projections were evenly spaced around the origin. All of Euler space was covered. In a real world example, the experimental projections are usually either randomly distributed or heavily biased and clustered towards one orientation. Ideally, an even, random distribution is prefered. However, this is rarely the case. When experimental projections are clustered, some views will be over represented over others. This results in a skewed final reconstruction. figure 6 is such an example of a biased reconstruction. The original model is the same as in figure 3, but the back projection was only limited to half the projection data set.

REFINING THE MODEL

The reference model obtained in the previous step is often further refined to increase the visible detail. A common method for refinement has been termed projection matching or angular refinement. This iterative process is used to better define the angular position of each experimental projection.

Projection Matching and Angular Refinement

In random conical tilt, images were assigned angular positions through rotational alignment and tilt-angles. As described earlier, particles are first classified according to their orientation on the carbon film. From each different class, a three-dimensional preliminary model is constructed. To improve the output, those preliminary models from each class that have a high degree of similarity are merged. In theory, these models corresponded to groups of the same molecule just viewed from different orientations. If a model does not sufficiently match the other models, it is discarded.

Once all the good random conical tilt models (and their corresponding particle data sets) have been merged, iterative angular refinement is used to improve the model's resolution. Equidistant projections are first generated from the merged model. A good angular separation distance used for the first run is 15°.

In this case, 89 reference projections are generated that cover the entire Euler sphere. The entire particle data set (whether the old random conical tilt experimental particles, or new untilted experimental particles, or both) is then cross correlated to each reference projection. A correlation coefficient is generated between each experimental particle and reference projection. For each individual experimental particle, it is matched to the reference projection that gave the highest correlation coefficient. Therefore, it is assumed that this particle matches the Euler angles (θ, φ, ψ) of the reference projection.

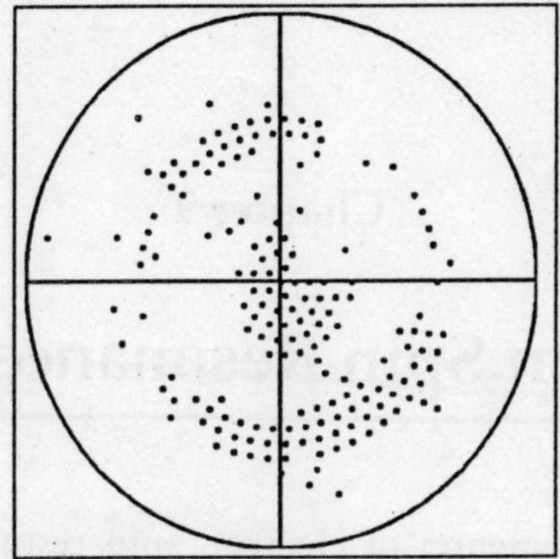

Fig. Angular Plot, Showing the Distribution of Particles Relative to a Reference Volume. Each dot Marks a Reference Pprojection that has a Matched Experimental Particle.

Figure is an angular distribution plot showing the spread of particles in angular space around the reference model. The θ and φ axes are indicated. Each mark inside the circle indicates the position of a matched projection. The dataset that generated this distribution plot consisted only of one class of particles and their corresponding tilt pairs.

The cluster of dots near the centre corresponds to the untilted particles, while the surrounding ring represents the tilted particles. Ideally, one would want the distribution plot to be completely filled, indicating that all possible views of the specimen have been imaged.

Now that each experimental projection has been matched with three Euler angles, a new model is generated via back projection. Reference projections are again generated from this model (though usually at a tighter angular distance such as 5°). Experimental particles are matched to the new reference projections and a new model is generated. This procedure is repeated until the reconstructed model does not improve during iterations.

Chapter 9

Electron Spin Resonance (ESR)

The phenomenon of electron spin resonance (ESR) is based on the fact that an electron is a charged particle which spins around its axis and this causes it to act like a tiny bar magnet. In technical language we say that it has a magnetic moment, the value of which is called the Bohr magneton.

If an external magnetic field Be is impressed on the system, the electron will align itself with the direction of this field and process around this axis. This behaviour is analogous to that of a spinning top in the earth's gravitational field. Increasing the applied magnetic field will induce the electron to process faster and acquire more energy of motion called kinetic energy. In practice, the magnetic field will divide the electrons into two groups. In one group the magnetic moments of the electrons are aligned with the magnetic field, while in the other group the magnetic moments are aligned opposite or antiparallel to this external field.

Classically the needle of a compass points north, but if disturbed it could point in any direction. At the microscopic level of the electron, classical mechanics is replaced by quantum mechanics and one of the consequences is that the electron can only point either in the same direction as the external magnetic field or opposite to it. If perturbed only will point in one of the two directions but not in between. These two possible orientations in the applied field correspond to the projections Ms = ±1/2 along the magnetic field direction where Ms is a dimensionless number used to designate the orientation. Each orientation is associated with a different energy, the one with the spins antiparallel to the external field (Ms = -1/2) being the lower energy state. If a second weaker alternating magnetic field B_1 oscillating at a microwave frequency is now applied at right angles to the main field B_0, then the electron can be "tipped" over when the microwave frequency is equal to the precession frequency. Another way to describe the phenomenon of ESR is to say that the quanta of the incident microwaves induce transitions between the two states of the unpaired electron. When the energy hvof these quanta coincides with the energy level separation :

$$\Delta E = E_{1/2} + E_{-1/2}$$

between the two states then resonance absorption of energy takes place.

The incoming radiation hv absorbed by the electrons in the lower energy level will induce these electrons to jump into the higher energy state. The incoming radiation is, however, also absorbed by the electrons in the higher energy level causing them to jump down to the lower level, a phenomenon called stimulated emission. Since the coefficients of absorption and stimulated emission are equal, no net value would be observed if the spin population were equally distributed between these two levels. In general, however, no, the population of the groundstate, exceeds n_2, the population of the excited state, and a net absorption of microwave radiation takes place. The population ratio of these two states at the temperature T can in most cases be described by the Boltzmann distribution

$$n_1 / n_2 = e^{-\Delta E / kT}$$

where k is a universal constant of nature called Boltzman's constant. A material containing atomic magnetic moments satisfying this Boltzmann distribution is called paramagnetic. Since E = hv at resonance the sensitivity of this technique is improved by using a high applied frequency. In practice, the most common frequencies in ESR are those in the well-developed radar wavebands. In addition, low operating temperatures are also beneficial for signal enhancement.

In most substances chemical bonding results in the pairing of the electrons which are transferred from one atom to another atom to form an ionic bond; or are shared between atoms to form a covalent bond so these materials are not magnetic. However, in a paramagnetic substance, i. e. one which contains unpaired electrons, resonance occurs at definite values of the applied magnetic field and incident microwave radiation.

At first glance one would expect the resonance spectrum of an unpaired electron to be always the same, but it is not because the magnetic behaviour of the electron is modified by its surroundings. This modification of the resonance by the surroundings permits us to learn something about the structure of the material under study.

In a typical experiment the sample is placed in a resonant cavity which has a high quality factor or high Q. At a fixed microwave frequency o, and the magnetic field is varied until resonance occurs at the value Be given by

$$hv = g\beta B_0$$

In this equation g is the "g-factor" which is a dimensionless constant and, β is the unit magnetic moment of the spinning electron called the Bohr magneton.

In ESR it is customary to measure the g-factor. Some transition ions such as cobalt (Co2+) and copper (Cu2+) have g-values which differ significantly form g = 2, but free radicals have values that are fairly close to g = 2.0023 which characterizes a free electron.

Most ESR spectra are more complex than is implied by equation 2. The resonant absorption is not infinitely narrow, which would be the case if absorption occurs at a precise value of the applied magnetic field. The actual line width ranges from a few milligauss for free radicals in solution to over 1000 gauss for some transition metal compounds in the solid state.

The reason for the finite line width ΔB is that unpaired electrons not only interact with the externally applied magnetic field but also interact in a more or less random manner with the magnetic fields in their neighborhood. By observing this line width and line intensity, one can obtain information about the spin environment. Electron spin exchange between identical and non-identical molecules, chemical exchange between the paramagnetic molecule and its environment, and the interaction of nearby molecules having unpaired spins are some examples of environmental effects which can influence line width and intensity in the ESR spectrum.

An observed spectrum sometimes contains several lines referred to as hyperfine structure arising from the electrons interacting with nuclear spins denoted by I. The electronic spin of a transition metal ion usually interacts with its own nuclear spin, and in aromatic free radicals, the unpaired electron circulates among several atoms; and the resultant hyperfine structure is the result of the interaction of this electronic spin with several atoms such as hydrogen with nuclear spins. An example is the ethyl radical :

```
     H     H
     |     |
H —  C  —  C  •
     |     |
     H     H
```

in which the unpaired electron (•) interacts with the nuclear spins of two hydrogen atoms on the same carbon atom and also with the three hydrogens of the adjacent methyl group (CH_3). The energy levels of an unpaired electron interacting with two nuclear spins is shown in figure 1.

The simplest case is one in which there is no hyperfine interaction I=0). By placing the unpaired electron in a magnetic field, the number of energy levels is increased from one (E = Eo) to two ($E = E_0 \pm 1/2 g\beta H$), as shown in the left portion of figure 1.

The centre and right side of the figure show cases where two nuclear spins with I=1/2 interact with the electron. A measure for the amount of interaction is given by the hyperfine coupling constant, A_i

Each nuclear spin I_1 and I_2 splits each original level into two levels. Hence, for A1≠ A2, the original two levels are split into 8 levels, as shown. Although there are 8 levels, only 4 transitions are possible because of the following selection rules.

$$\Delta M_S = \pm 1$$
$$\Delta M_I = \pm 0$$

where M_s is the electron spin quantum number and M_I is the nuclear spin quantum number. The hyperfine coupling constant Al with the nuclear species, and hyperfine splittings of the order of 10 millitesla or more can be observed. In organic compounds the electron may interact with several nuclei, and the hyperfine interactions may be as low as a few microtesla. The technique of ESR has been used to study many types of systems and we will comment on a few of them.

- Biological Systems : ESR has been applied quite extensively to biological systems. One can follow the variations that occur under changing environmental conditions by monitoring the intensity of a free radical signal. For example, the presence of free radicals has been studied in healthy and diseased tissue. If a transition metal ion is present, as in case of the Fe ions of hemoglobin, then changes in its valence state may be studied by ESR. Early concrete evidence for the role of free radicals in photosynthesis, was provided by the observation of a sharp ESR resonance line. When incident light was turned off the resonance soon weakened or disappeared.

An inconvenience associated with the analysis of biological samples is the presence of water which produces high dielectric losses, making it necessary to use a flat quartz sample cell.

Some typical systems which have been studied by ESR are hemoglobin, nucleic acids, enzymes, irradiated chloroplasts, riboflavin (before and after UV irradiation), and carcinogens.

- Chemical Systems : A number of chemical substances such as synthetic polymers and rubber contain free radicals. The nature of the free radical depends upon the method of synthesis and on the history of the substance. Recent refinements in instrumentation have permitted the detection of free radical intermediates in chemical reactions.

Typical chemical systems that have been studied by ESR are polymers, catalysts, rubber, along-lived free radical intermediates, charred carbon, and chemical complexes with transition metals.

- Conduction Electrons : Conduction electrons have been detected by both conventional electron spin resonance methods, and by the related cyclotron resonance technique. The latter employs and ESR spectrometer and the sample is located in a region of strong microwave electric field strength. In the usual ESR arrangement, on the other hand, the sample is placed at a position of strong microwave magnetic field strength. Conduction electrons have been detected in solutions of alkali metals in liquid ammonia, alkaline earth metals (fine powders), alloys (e. g., small amounts of paramagnetic metal alloyed with another metal), non-resonant absorption of microwaves by superconductors, and graphite.
- Free Radicals : A free radical is a compound which contains and unpaired spin, such as the methyl radical CH3 produced through the breakup of methane

$$CH_4 \rightarrow CH_3 + H\bullet$$

where both the hydrogen atom and the methyl radical are electrically neutral. A charged free radical or radical ion is a neutral molecule which has gained or lost an electron

$$C_6H_6 + e \rightarrow C_6H_6^{\,-} \text{(cation)}$$

$$C_6H_6 \rightarrow e + C_6H_6^{\,+} \text{ (anion)}$$

Free radicals have been observed in gaseous, liquid, and solid systems. They are sometimes stable, but can also be shod-lived intermediates in chemical reactions.

Free radicals and radical ions ordinarily have g-factors close to the free electron value of 2.0023 (e. g., for the stable free radical α,α' diphenylα-picryl-hydrazyl, referred to as DPPH, g = 2.0036). In low-viscosity solutions, radicals exhibit hyperfine patterns with a typical overall spread of about 2.5 millitesla. The scrupulous removal of oxygen often reveals hitherto unresolved structure. In a high-concentration solid a single exchange narrowed resonance appears ($\Delta B \sim 0.3$ millitesla for DPPH prepared from benzene solution), whereas DPPH in a dilute oxygen free solution of the solvent tetrahydrofurane produces a spectrum with more than 100 hyperfine components.. In irradiated single crystals the free radicals may have strongly directionally dependent or anisotropic hyperfine interactions and slightly anisotropic g-factors.

Radical ions of many organic compounds can be produced in an electrolytic cell which is usually a flat quartz cell with a mercury pool cathode and a platinum anode. This electrolytic cell may be mounted in a flat measuring cell located in the resonance cavity.

When the applied voltage in the electrolytic system is increased, the current will first increase, but soon levels off to a plateau. Radical ions are formed in this plateau region. Radical formation can sometimes be observed visually because of colour changes in the solution. Since oxygen is also paramagnetic, dissolved air must be scrupulously removed prior to the experiment.

The best method is to use the freeze-pump-thaw technique where the sample is first frozen and then connected to a high vacuum source. After closing off the vacuum pump, the sample is melted and refrozen. Several cycles are sufficient to remove the oxygen.

- *Irradiated Substances:* A considerable amount of work has been done on radiation induced free radicals in organic compounds and paramagnetic colour centers in ionic solids. Most irradiations are carried out with x-rays, gamma-rays, or electrons whose energies far exceed chemical bond energies, although paramagnetic spins can also be produced by less energetic ultraviolet light or neutrons. Most of these spectra are obtained after the sample is irradiated, and many paramagnetic centers are found to be sufficiently long-lived to warrant such a procedure. More sophisticated experimental techniques entail simultaneous irradiation and ESR detection. Low-temperature irradiation and detection can reveal the presence of new centers which can be studied at gradually increasing temperatures to elucidate the kinetics of their recombination.

- *Naturally Occurring Substances:* Most of the systems studied by ESR are synthetic or man-made. Nevertheless, from the beginning of the field, various naturally occurring substances have been studied, such as: (1) minerals with transition elements [e. g., ruby (Cr/Al2O3), dolomite Mn/(Ca, Mg(Co3)]; (2) minerals with defects (e. g., quartz); (3) hemoglobin (Fe); (4) petroleum; (5) coal; (6) rubber; and (7) various biological systems.
- *Spin labels:* There are a number of free radicals called spin labels which can attach themselves to particular sites in biological systems and produce spectra which provide information on changes in the chemical and physical characteristics in the neighborhood of the site. An example of a spin label has its unpaired electron on the NO group of a nitroxyl compound such as 2, 2, 6, 6-tetramethyl-4-piperadone-1-oxyl

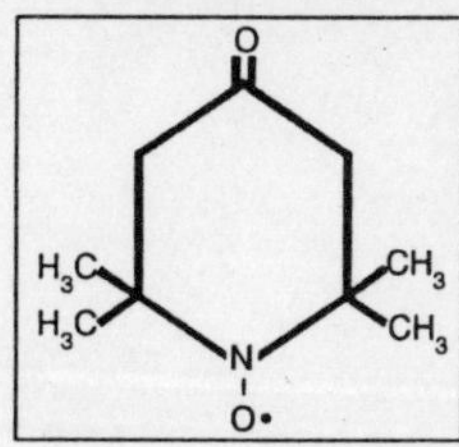

The unpaired electron shown on the lower oxygen, spends time on and strongly interacts with the I = 1 nuclear spin of the nitrogen atom to produce a 3-line hyperfine spectrum. When the solvent has a low viscosity (flows very freely) the radical tumbles rapidly to average out the anisotropies, and a spectrum of three narrow lines is obtained. At lower temperatures the viscosity increases and the solvent materials flows very slowly so the nitrogen radical can no longer tumble rapidly, and a smeared out spectrum is produced.

When the spin label attaches itself to a particular site in a biological molecule its spectrum reveals the extent to which free motion or very restricted motion occurs at the site.

Different spin labels have been synthesized which attach themselves to very specific sites. For example the spin label n-doxylstearic has the nitroxide doxyl group attached to different positions n between 5 and 16 on the long stearic acid molecule.

This spin label can be inserted into a cell membrane so each value of n corresponds to a different depth in the membrane, and a sequence of labels provides information on the activities of the membrane at various depths below the surface.

Chapter 10

Frequency Dependence in Free Space Propagation

It seems to be fairly common knowledge, even among practicing professionals, that the efficiency of propagation of wireless signals is frequency dependent. Generally it is believed that lower frequencies are desirable since pathloss effects will be less than they would be at higher frequencies. As evidence of this, the *Friis Transmission Equation* is often cited, the general form of which is usually written as:

$$P_r = P_t \, G_t \, G_r \, (\lambda / 4\eth d)^2 \quad (1)$$

where the $(\lambda/4\eth d)^2$ term is sometimes referred to as the *path loss* or *free space loss*. The following definitions are assumed:

P_r - The received signal power.

P_t - The transmitted signal power.

G_r - The gain of the receiving antenna.

G_t - The gain of the transmitting antenna.

λ - The wavelength of the carrier (i.e., the centre frequency of the radiated signal)

d - The distance between the transmitting and receiving antennas.

It is evident from equation (1) that, all other variables held constant, as the transmission frequency is reduced, the wavelength, λ, increases. The received power, P_r, then goes up by the *square* of the wavelength! *Clearly it is always desirable to use as low a frequency as possible, correct?*

PROPAGATION

One source of confusion is what is meant by "propagation". For the purposes of this discussion, and in order to fully illuminate the root of the issue, it is good to get a better understanding of what is meant by "channel effects" and "propagation effects". A "channel" is generally, and I'll take a little license here, the medium through which a given set of information is conveyed. In a wireless system this usually includes what I call "the propagation channel" as well as "the radio channel".

The "propagation channel" includes things like path loss, multipath reflection effects, multipath fading, Doppler (which can also be induced by moving reflectors in the environment, even if the radio terminals are stationary), fading due to material penetration (e.g., wall penetration, cloud penetration, etc.), etc., etc. For the most part, I think of the "propagation channel" as what happens to the signal between the antennas, and for applications like mobile radio or Non-Line-of-Sight links "propagation channel" effects can dominate the impairments and distortion that the demodulator must handle in order to recover the transmitted information.

The terminology of "channel effects" is a bit broader, and includes "propagation effects" but also adds what I call "radio channel" effects, or the effects of the antennas, signal cabling, filter distortions, amplifier distortions and nonlinearities, amplifier noise, mixing distortions, quantization noise, etc., etc. The idea is to capture pretty much any distortion or impairment that happens to the signal from the point that it is "modulated" into a signal and the point at which is it "demodulated" back into the estimate of the original transmitted information. This includes imperfections in signal processing and radio components.

Other impairment effects, like interference from other signals, may be lumped into "propagation effects" even though they do not affect the physical propagation of the electromagnetic wave carrying the signal. Even by my definition, that "propagation effects" are what happens between the antennas, one would have to include interference since an interference source could easily be, and often is, between the transmit and receive antennas.

Another important issue is whether or not the antenna itself is included when discussing "propagation", and this winds up being a key point, if not *the* key point. Antenna response and antenna design generally is highly frequency dependent, but the propagation of an electromagnetic wave in free space is not. So whether or not one means to include antennas in a discussion regarding "propagation" makes all the difference in the world.

So it's not hard to see why there is often disagreement or just misunderstanding regarding where the "radio effects" start and stop and where the "propagation effects" take over, and whether or not "propagation" includes antennas or not, or interference or not, blah blah blah.

The immediate point is not to sort all that out, but just to understand that there are significant sources of the confusion and that they're pretty common. It is very rare that a speaker, or standard, or proposal, or text book[iii], clarifies this distinction. Many times this may be because that level of detail is inappropriate for the particular discussion, or the clarification too peripheral to the main point, or because the source is also unaware of the confusion or distinction.

Gaining Knowledge About Antennas

Anybody who's spent much time with an antenna designer knows that there's a fair amount of art involved in the process. There are important fine details and mysterious field effects, material properties, planetary alignments and other things beyond the understanding of mere mortals that have significant influences on antenna behaviour. Fortunately, for the purposes of our discussion here, we can leave the padlocks safely secure on the doors that separate the antenna designers from the rest of society. The aspects of antenna behaviour that come into play for understanding the fundamentals of frequency dependence are pretty basic, enough so that I think even I can take a crack at explaining them well enough to get to the point.

Consider a point radiator that emanates a certain amount

of power, P, equally in all directions, i.e., the ideal omnidirectional antenna. The area covered by the antenna radiation in free space at a distance, d, from the radiator is just the area of a sphere, A_{sphere} = 4ðd^2. As the distance from the radiator increases, the area covered increases with the square of the distance.

We can say that the power flux density decreases with the square of the distance. A common intuitive example to help imagine this is the inflation of a balloon, perhaps with a small square drawn on the surface. As the balloon is inflated the skin of the balloon gets thinner as it has to cover more area, which is demonstrated by the increasing area of the square as the radius of the balloon increases. The power flux density of the transmitted signal likewise grows thinner with distance, and the recoverable power within a given area drops by the square of the distance.

All antennas have an area metric called the Effective Area or Effective Aperture, usually denoted as A_e, which describes the antenna's ability to collect power. Continuing with the radiating sphere example, since the power flux density from an omnidirectional radiator can be computed from the distance, the Effective Area of an antenna can be determined by measuring the amount of power collected at given distance from an omnidirectional radiation of P power, or in a given power flux density. So, in simple terms, A_e is the ratio of the collected power to the power flux density incident upon it.

The effective area may or may not have much to do with or correlate well with the physical dimensions of the antenna. It is merely an indicator of the power collection ability of the device.

So the amount of power collected at a receiving antenna with effective area A_{er} from an ideal omnidirectional radiator transmitting P_t power is, strictly from the geometric relationship:

$$P_r = P_t A_{er} / 4\pi d^2$$

Note that there is no frequency dependence in the relationship shown. The only variables that affect the amount of power received are the propagation distance, d, the transmitted power, P_t, and the effective area of the receiving antenna.

Equation can be generalized for a non-omnidirectional transmit antenna by substituting the Effective Isotropic Radiated Power (EIRP) for the total transmit omnidirectional transmit power, P_t. Since the basic concept of EIRP is to quantify the power flux density independent of the gain or configuration of the transmit antenna, this is a useful step in further generalizing the relationship, so that

$$P_r = EIRP\ A_{er} / 4\pi d^2$$

Again, we note that there is still no frequency dependence expressed in equation regardless of what type of antennas are used. Since equations and suggest that the propagation of the electromagnetic waves have no frequency dependence, there is clearly some factor in play that introduces the wavelength term, λ, into equation. To understand this we need only to look closely at the common expression for antenna gain, specifically

$$G = 4\pi A_e/\lambda^2$$

Equation can be found in many texts on communications, electromagnetics, or antennas. It is given here without background for brevity.

There are several things we can learn from equation, and we don't even need to delve far into the details of antenna theory or design. Let's consider breaking quation into two separate cases: omnidirectional radiators, specifically monopole or dipole antennas, and directive reflecting antennas, commonly referred to as "dish" antennas.

In the omnidirectional, dipole or monopole, case, we know that the size of the antenna needs to be matched to the frequency of operation for maximum efficiency. Generally a stub element in a monopole antenna is cut to a quarter-wavelength of the desired operating frequency, and a dipole consists of two such quarter-wave stubs butted up against each other as shown in figure.

Since this means that the physical size of the antenna gets smaller when optimized for a particular higher frequency, this has an effect on the antenna's ability to collect energy. Setting

G = 1 in equation, as would be the case for an omnidirectional radiator, we see that

$$A_e = \lambda^2 / 4\pi$$

which reveals that, sure enough, the effective area, A_e, increases with the square of the wavelength, so that lower frequencies provide a substantial advantage in energy collection. This is not difficult to understand intuitively; dipole or monopole antennas get physically larger when sized for lower frequencies. It is this characteristic, that an omnidirectional stick antenna's effective area increases when physically dimensioned for lower frequencies, that misleads many people into believing that "propagation" is frequency dependent.

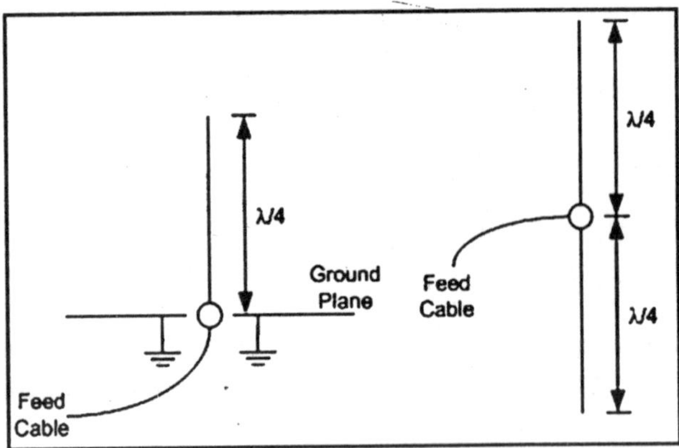

Fig. A Monopole (left) and Dipole (right) Antenna, with the Quarter-wavelength Dimensions Shown.

We should be careful to understand, though, that this effect relates to stick antennas that are physically dimensioned to a particular frequency, e.g., monopoles cut to quarter-wavelengths. A given antenna will exhibit frequency selectivity in that it will have a bandpass characteristic related to the resonating efficiency of the particular design. Generally dipole or monopole antennas have well-behaved bandpass characteristics that can cover bands of interest with reasonable efficiency. For any given stick antenna, then, the only frequency dependence has to do with the particular bandpass frequency selectivity of the design, and one expects a monopole or dipole to have a peak response matched to the quarter-wavelength related to the physical size of the antenna. figure 2 illustrates the "frequency dependence" shown in compared to typical responses of stick antennas at various frequencies.

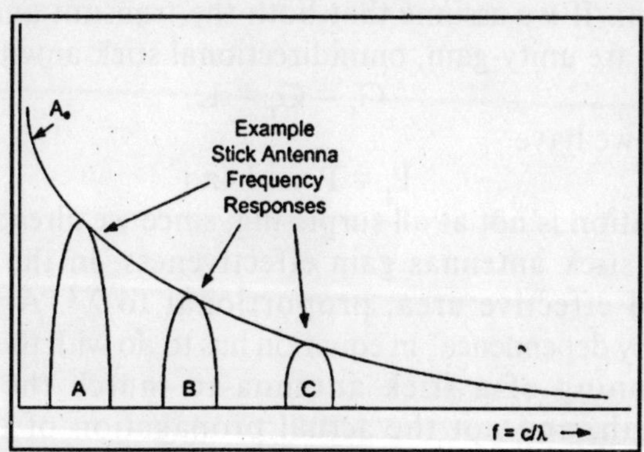

Fig. The Effective Aarea, A_e, of a Stick Antenna, with the Dimensions Shown in figure, Increases with the Square of the Wavelength. Individual Antennas Cut to Specific Frequencies, Like those Shown as A, B, and C, will Generally have Bandpass Frequency Selectivity.

For the reflecting dish antenna case the physical size of the antenna also plays a revealing part in illuminating the mystery of frequency dependent propagation. We can see in equation that if we desire to build a high-gain (i.e., high-G) antenna, we can increase A_e or we can decrease the wavelength, λ. For a reflecting dish of a given size, A_e will be fixed by the dimensions of the device, but the gain will *increase* with frequency. Again we see a frequency dependence related to the antenna but in the opposite sense of the omnidirectional stick antenna case. Instead of a loss with increasing frequency, we see antenna gain improving with the square of the frequency.

Closing the Link Between Antennas and Frequency Dependence

Now that we understand that the type of antenna has a significant effect on the "frequency dependence" of propagation, let's take another look at the Friis Transmission equation. It is not difficult to start from all equations and derive equation. Along the way one might stumble across the point which we will now attempt to make.

The general form of the equation shown in equation is convenient when the gains of the transmit and receive antennas are known. If we assume that both the transmit and receive antennas are unity-gain, omnidirectional stick antennas, i.e.,

$$G_t = G_r = 1,$$

then we have

$$P_r = P_t \lambda^2 / 4\pi d^2$$

Equation is not at all surprising, since we already sorted out that stick antennas gain effectiveness, in the form of increased effective area, proportional to λ^2. Again, the "frequency dependence" in equation has to do with the physical dimensioning of a stick antenna to match the desired wavelength, and not the actual propagation of the wave through space. If someone were to apply equation to a link with a carrier at a particular wavelength but antennas dimensioned for a different frequency, they might spend a fair amount of time trying to reconcile the discrepancy if they were unaware of the basis for the wavelength term in the equation.

Likewise if we have a dish antenna at each end of the link, both with fixed dimensions and therefore fixed effective areas, we could substitute equation into equation and write

$$P_r = P_t A_{et} A_{er} / \lambda^2 d^2$$

Again, we experience no surprise that λ has moved from the numerator in equation to the denominator in equation, since we've already determined that reflective dish antennas of fixed area experience a gain *improvement* proportional to the square of the frequency. We also know that equation is not in contradiction to equation and is essentially a recognizable form of equation. We can now anticipate the inevitable combination of fixed gain omnidirectional and fixed area reflecting antenna and write

$$P_r = P_t G_t A_{er} / 4\pi d^2$$

or the simple rearrangement switching the forms of the transmit and receive antennas

$$P_r = P_t A_{et} G_r / 4\pi d^2$$

As expected, in these cases the "frequency dependent" characteristics of the antennas cancel and there is no longer a frequency-related term in equations at all. This helps demonstrate to those who may remain skeptical that the wavelength term in the Friis Transmission equation has to do with the antennas and not the actual propagation of the wave through space.

Meaning

The Friis equation, in the various forms we are now familiar with, addresses link behaviour in free space. We now understand that free space propagation between the antennas is completely independent of frequency. An important caveat, though, is that many links have something other than free space between the antennas. The propagation of a wave from a fixed broadcast tower to a receiver inside a house or building must penetrate the roof, walls, windows, doors, tree leaves, or whatever materials lie in the path between the transmitter and receiver. It is very often the case that a wave must penetrate various materials on its way to whatever receiving antennas it may encounter.

It is well known and generally true that material penetration is frequency dependent, and generally attenuation increases with frequency. That is, the higher the frequency of the wave the more opaque that materials appear to be. Since the type and thickness of the material makes a difference in the amount of attenuation experienced, consideration of penetration losses is a much more complicated endeavor than the free space analysis that we've just done. So while material penetration loss generally increases with frequency, specific cases and scenarios require detailed analysis suitable for the particular application.

Some materials exhibit substantial attenuation properties at particular frequencies and less attenuation at frequencies above or below their resonance band. A notable example is the oxygen absorption bands around 60GHz. The range of radio waves at these frequencies is decreased significantly by the energy absorption of atmospheric oxygen molecules.

Summary, Conclusion, and Chest-Beating

It is apparent from careful study of the Friis Transmission Equation that the often-mentioned "frequency-dependent propagation loss" of radio waves is really an antenna effect and not a wave propagation effect. The propagation of a radio wave or photon through free space is unaffected by its frequency.

Stick antennas, that is, dipole or monopole antennas, are larger at lower frequencies and therefore have more effective area for energy collection. Reflective antennas, that is dish antennas, of a given size have higher gain with decreasing wavelength. Systems that use stick antennas therefore benefit from lower frequencies while dish-antenna systems will attain higher system gain at higher frequencies.

Remember to use this knowledge for good and not for evil, although it should be possible to impress smug colleagues if the topic ever comes up.

Chapter 11

Infrared

Infrared (IR) radiation is electromagnetic radiation whose wavelength is longer than that of visible light (400-700nm), but shorter than that of terahertz radiation (3-300μm) and microwaves (~30,000um). Infrared radiation spans roughly three orders of magnitude (750nm and 1000μm).

Direct sunlight has a luminous efficacy of about 93 lumens per watt of radiant flux, which includes infrared (47% share of the spectrum), visible (46%), and ultra-violet (only 7%) light. Bright sunlight provides luminance of approximately 100,000 candela per square meter at the Earth's surface.

Infrared imaging is used extensively for both military and civilian purposes. Military applications include target acquisition, surveillance, night vision, homing and tracking. Non-military uses include thermal efficiency analysis, remote temperature sensing, short-ranged wireless communication, spectroscopy, and weather forecasting. Infrared astronomy uses sensor-equipped telescopes to penetrate dusty regions of space, such as molecular clouds; detect cool objects such as planets, and to view highly red-shifted objects from the early days of the universe.

Humans at normal body temperature radiate chiefly at wavelengths around 10μm (micrometers).

At the atomic level, infrared energy elicits vibrational modes in a molecule through a change in the dipole moment, making it a useful frequency range for study of these energy states. Infrared spectroscopy examines absorption and transmission of photons in the infrared energy range, based on their frequency and intensity.

DIFFERENT REGIONS IN THE INFRARED

Objects generally emit infrared radiation across a spectrum of wavelengths, but only a specific region of the spectrum is of interest because sensors are usually designed only to collect radiation within a specific bandwidth. As a result, the infrared band is often subdivided into smaller sections.

The International Commission on Illumination (CIE) recommended the division of optical radiation into the following three bands:

- IR-A: 700nm–1400nm
- IR-B: 1400nm–3000nm
- IR-C: 3000nm–1mm

A commonly used sub-division scheme is:[citation needed]

- Near-infrared (NIR, IR-A *DIN*): 0.75-1.4μm in wavelength, defined by the water absorption, and commonly used in fibre optic telecommunication because of low attenuation losses in the SiO_2 glass (silica) medium. Image intensifiers are sensitive to this area of the spectrum. Examples include night vision devices such as night vision goggles.
- Short-wavelength infrared (SWIR, IR-B *DIN*): 1.4-3μm, water absorption increases significantly at 1,450nm. The 1,530 to 1,560nm range is the dominant spectral region for long-distance telecommunications
- Mid-wavelength infrared (MWIR, IR-C *DIN*) also called intermediate infrared (IIR): 3-8μm. In guided missile technology the 3-5μm portion of this band is the atmospheric window in which the homing heads of passive IR 'heat seeking' missiles are designed to work, homing on to the IR signature of the target aircraft, typically the jet engine exhaust plume.
- Long-wavelength infrared (LWIR, IR-C *DIN*): 8–15μm. This is the "thermal imaging" region, in which sensors can obtain a completely passive picture of the outside world based on thermal emissions only and requiring no external light or thermal source such as the sun, moon or infrared illuminator.

Forward-looking infrared (FLIR) systems use this area of the spectrum. Sometimes also called the "far infrared."

- Far infrared (FIR): 15-1,000μm (see also far infrared laser)

NIR and SWIR is sometimes called *reflected infrared* while MWIR and LWIR is sometimes referred to as *thermal infrared*. Due to the nature of the blackbody radiation curves, typical 'hot' objects, such as exhaust pipes, often appear brighter in the MW compared to the same object viewed in the LW.

Astronomers typically divide the infrared spectrum as follows:

- near: (0.7-1) to 5 μm
- mid: 5 to (25-40) μm
- long: (25-40) to (200-350) μm

These divisions are not precise and can vary depending on the publication. The three regions are used for observation of different temperature ranges, and hence different environments in space.

A third scheme divides up the band based on the response of various detectors:

- Near infrared (NIR): from 0.7 to 1.0 micrometers (from the approximate end of the response of the human eye to that of silicon)
- Short-wave infrared (SWIR): 1.0 to 3 micrometers (from the cut off of silicon to that of the MWIR atmospheric window. InGaAs covers to about 1.8 micrometers; the less sensitive lead salts cover this region
- Mid-wave infrared (MWIR): 3 to 5 micrometers (defined by the atmospheric window and covered by Indium antimonide [InSb] and HgCdTe and partially by lead selenide [PbSe])
- Long-wave infrared (LWIR): 8 to 12, or 7 to 14 micrometers: the atmospheric window (Covered by HgCdTe and microbolometers)
- Very-long wave infrared (VLWIR): 12 to about 30 micrometers, covered by doped silicon

These divisions are justified by the different human response to this radiation: near infrared is the region closest in wavelength to the radiation detectable by the human eye, mid and far infrared are progressively further from the visible regime. Other definitions follow different physical mechanisms (emission peaks, vs. bands, water absorption) and the newest follow technical reasons (The common silicon detectors are sensitive to about 1,050nm, while InGaAs' sensitivity starts around 950nm and ends between 1,700 and 2,600nm, depending on the specific configuration). Unfortunately, international standards for these specifications are not currently available.

The boundary between visible and infrared light is not precisely defined. The human eye is markedly less sensitive to light above 700 nm wavelength, so shorter frequencies make insignificant contributions to scenes illuminated by common light sources. But particularly intense light (e.g., from lasers, or from bright daylight with the visible light removed by colored gels) can be detected up to approximately 780 nm, and will be perceived as red light. The onset of infrared is defined (according to different standards) at various values typically between 700 nm and 800 nm.

Telecommunication Bands in the Infrared

In optical communications, the part of the infrared spectrum that is used is divided into several bands based on availability of light sources, transmitting/absorbing materials (fibers) and detectors:

Band	Descriptor	Wavelength Range
O band	Original	1260–1360nm
E band	Extended	1360–1460nm
S band	Short wavelength	1460–1530nm
C band	Conventional	1530–1565nm
L band	Long wavelength	1565–1625nm
U band	Ultralong wavelength	1625–1675nm

The C-band is the dominant band for long-distance

telecommunication networks. The S and L bands are based on less well established technology, and are not as widely deployed.

Heat

Infrared radiation is popularly known as "heat" or sometimes "heat radiation", since many people attribute all radiant heating to infrared light and/or to all infrared radiation to being a result of heating. This is a widespread misconception, since light and electromagnetic waves of any frequency will heat surfaces that absorb them. Infrared light from the Sun only accounts for 49% of the heating of the Earth, with the rest being caused by visible light that is absorbed then re-radiated at longer wavelengths. Visible light or ultraviolet-emitting lasers can char paper and incandescently hot objects emit visible radiation. It is true that objects at room temperature will emit radiation mostly concentrated in the 8 to 12 micrometer band, but this is not distinct from the emission of visible light by incandescent objects and ultraviolet by even hotter objects (see black body and Wien's displacement law).

Heat is energy in transient form that flows due to temperature difference. Unlike heat transmitted by thermal conduction or thermal convection, radiation can propagate through a vacuum.

The concept of emissivity is important in understanding the infrared emissions of objects. This is a property of a surface which describes how its thermal emissions deviate from the ideal of a black body. To further explain, two objects at the same physical temperature will not 'appear' the same temperature in an infrared image if they have differing emissivities.

APPLICATIONS

Infrared Filters

Infrared (IR) filters can be made from many different materials. One type is made of polysulphone plastic that blocks over 99% of the visible light spectrum from "white" light sources such as incandescent filament bulbs. Infrared filters allow a maximum of infrared output while maintaining extreme covertness. Currently in use around the world, infrared filters are used in Military, Law Enforcement, Industrial and Commercial applications. The unique makeup of the plastic allows for maximum durability and heat resistance. IR filters provide a more cost effective and time efficient solution over the standard bulb replacement alternative. All generations of night vision devices are greatly enhanced with the use of IR filters.

Night Vision

Infrared is used in night vision equipment when there is insufficient visible light to see. Night vision devices operate through a process involving the conversion of ambient light photons into electrons which are then amplified by a chemical and electrical process and then converted back into visible light. Infrared light sources can be used to augment the available ambient light for conversion by night vision devices, increasing in-the-dark visibility without actually using a visible light source.

The use of infrared light and night vision devices should not be confused with thermal imaging which creates images based on differences in surface temperature by detecting infrared radiation (heat) that emanates from objects and their surrounding environment

Thermography

Infrared radiation can be used to remotely determine the temperature of objects (if the emissivity is known). This is termed thermography, or in the case of very hot objects in the NIR or visible it is termed pyrometry. Thermography (thermal imaging) is mainly used in military and industrial applications but the technology is reaching the public market in the form of infrared cameras on cars due to the massively reduced production costs.

Thermographic cameras detect radiation in the infrared

range of the electromagnetic spectrum (roughly 900–14,000 nanometers or 0.9–14 μm) and produce images of that radiation. Since infrared radiation is emitted by all objects based on their temperatures, according to the black body radiation law, thermography makes it possible to "see" one's environment with or without visible illumination. The amount of radiation emitted by an object increases with temperature, therefore thermography allows one to see variations in temperature (hence the name).

Other Imaging

In infrared photography, infrared filters are used to capture the near-infrared spectrum. Digital cameras often use infrared blockers. Cheaper digital cameras and camera phones have less effective filters and can "see" intense near-infrared, appearing as a bright purple-white colour.

This is especially pronounced when taking pictures of subjects near IR-bright areas (such as near a lamp), where the resulting infrared interference can wash out the image. There is also a technique called 'T-ray' imaging, which is imaging using far infrared or terahertz radiation.

Lack of bright sources makes terahertz photography technically more challenging than most other infrared imaging techniques. Recently T-ray imaging has been of considerable interest due to a number of new developments such as terahertz time-domain spectroscopy.

Tracking

Infrared tracking, also known as infrared homing, refers to a passive missile guidance system which uses the emission from a target of electromagnetic radiation in the infrared part of the spectrum to track it. Missiles which use infrared seeking are often referred to as "heat-seekers", since infrared (IR) is just below the visible spectrum of light in frequency and is radiated strongly by hot bodies. Many objects such as people, vehicle engines and aircraft generate and retain heat,and as such, are especially visible in the infra-red wavelengths of light compared to objects in the background.

Heating

Infrared radiation can be used as a deliberate heating source. For example it is used in infrared saunas to heat the occupants, and also to remove ice from the wings of aircraft (de-icing). FIR is also gaining popularity as a safe method of natural health care & physiotherapy. Far infrared thermomedic therapy garments use thermal technology to provide compressive support and healing warmth to assist symptom control for arthritis, injury & pain. Infrared can be used in cooking and heating food as it predominantly heats the opaque, absorbent objects, rather than the air around them.

Infrared heating is also becoming more popular in industrial manufacturing processes, e.g. curing of coatings, forming of plastics, annealing, plastic welding, print drying. In these applications, infrared heaters replace convection ovens and contact heating. Efficiency is achieved by matching the wavelength of the infrared heater to the absorption characteristics of the material.

Communications

IR data transmission is also employed in short-range communication among computer peripherals and personal digital assistants. These devices usually conform to standards published by IrDA, the Infrared Data Association. Remote controls and IrDA devices use infrared light-emitting diodes (LEDs) to emit infrared radiation which is focused by a plastic lens into a narrow beam. The beam is modulated, i.e. switched on and off, to encode the data. The receiver uses a silicon photodiode to convert the infrared radiation to an electric current. It responds only to the rapidly pulsing signal created by the transmitter, and filters out slowly changing infrared radiation from ambient light. Infrared communications are useful for indoor use in areas of high population density. IR does not penetrate walls and so does not interfere with other devices in adjoining rooms. Infrared is the most common way for remote controls to command appliances.

Free space optical communication using infrared lasers can be a relatively inexpensive way to install a communications link in an urban area operating at up to 4 gigabit/s, compared to the cost of burying fibre optic cable.

Infrared lasers are used to provide the light for optical fibre communications systems. Infrared light with a wavelength around 1,330 nm (least dispersion) or 1,550 nm (best transmission) are the best choices for standard silica fibers.

IR data transmission of encoded audio versions of printed signs is being researched as an aid for visually impaired people through the RIAS (Remote Infrared Audible Signage) project.

Spectroscopy

Infrared vibrational spectroscopy (see also near infrared spectroscopy) is a technique which can be used to identify molecules by analysis of their constituent bonds. Each chemical bond in a molecule vibrates at a frequency which is characteristic of that bond. A group of atoms in a molecule (e.g. CH_2) may have multiple modes of oscillation caused by the stretching and bending motions of the group as a whole. If an oscillation leads to a change in dipole in the molecule, then it will absorb a photon which has the same frequency. The vibrational frequencies of most molecules correspond to the frequencies of infrared light. Typically, the technique is used to study organic compounds using light radiation from 4000-400 cm^{-1}, the mid-infrared. A spectrum of all the frequencies of absorption in a sample is recorded. This can be used to gain information about the sample composition in terms of chemical groups present and also its purity (for example a wet sample will show a broad O-H absorption around 3200 cm^{-1}).

METEOROLOGY

Weather satellites equipped with scanning radiometers produce thermal or infrared images which can then enable a trained analyst to determine cloud heights and types, to calculate land and surface water temperatures, and to locate ocean surface features. The scanning is typically in the range 10.3-12.5 μm (IR4 and IR5 channels).

High, cold ice cloud such as Cirrus or Cumulonimbus

show up bright white, lower warmer cloud such as Stratus or Stratocumulus show up as grey with intermediate clouds shaded accordingly. Hot land surfaces will show up as dark grey or black. One disadvantage of infrared imagery is that low cloud such as stratus or fog can be a similar temperature to the surrounding land or sea surface and does not show up. However, using the difference in brightness of the IR4 channel (10.3-11.5 μm) and the near-infrared channel (1.58-1.64 μm), low cloud can be distinguished, producing a *fog* satellite picture. The main advantage of infrared is that images can be produced at night, allowing a continuous sequence of weather to be studied.

These infrared pictures can depict ocean eddies or vortices and map currents such as the Gulf Stream which are valuable to the shipping industry. Fishermen and farmers are interested in knowing land and water temperatures to protect their crops against frost or increase their catch from the sea. Even El Nipo phenomena can be spotted. Using colour-digitized techniques, the gray shaded thermal images can be converted to colour for easier identification of desired information.

Climatology

In the field of climatology, atmospheric infrared radiation is monitored to detect trends in the energy exchange between the earth and the atmosphere. These trends provide information on long term changes in the earth's climate. It is one of the primary parameters studied in research into global warming together with solar radiation.

A pyrgeometer is utilized in this field of research to perform continuous outdoor measurements. This is a broadband infrared radiometer with sensitivity for infrared radiation between approximately 4.5 μm and 50 μm.

Astronomy

Astronomers observe objects in the infrared portion of the electromagnetic spectrum using optical components, including mirrors, lenses and solid state digital detectors. For this reason it is classified as part of optical astronomy. To form an image, the components of an infrared telescope need to be carefully shielded from heat sources, and the detectors are chilled using liquid helium.

The sensitivity of Earth-based infrared telescopes is significantly limited by water vapour in the atmosphere, which absorbs a portion of the infrared radiation arriving from space outside of selected atmospheric windows. This limitation can be partially alleviated by placing the telescope observatory at a high altitude, or by carrying the telescope aloft with a balloon or an aircraft. Space telescopes do not suffer from this handicap, and so outer space is considered the ideal location for infrared astronomy.

The infrared portion of the spectrum has several useful benefits for astronomers. Cold, dark molecular clouds of gas and dust in our galaxy will glow with radiated heat as they are irradiated by imbedded stars. Infrared can also be used to detect protostars before they begin to emit visible light. Stars emit a smaller portion of their energy in the infrared spectrum, so nearby cool objects such as planets can be more readily detected. (In the visible light spectrum, the glare from the star will drown out the reflected light from a planet.)

Infrared light is also useful for observing the cores of active galaxies which are often cloaked in gas and dust. Distant galaxies with a high redshift will have the peak portion of their spectrum shifted toward longer wavelengths, so they are more readily observed in the infrared.

Art History

Infra-red (as art historians call them) reflectogramsare taken of paintings to reveal underlying layers, in particular the underdrawing or outline drawn by the artist as a guide. This often uses carbon black which shows up well in reflectograms, so long as it has not also been used in the ground underlying the whole painting. Art historians are looking to see if the visible layers of paint differ from the under-drawing or layers in between - such alterations are called pentimenti when made by the original artist. This is very useful information in deciding whether a painting is the prime version by the original artist or a copy, and whether it has been altered by over-enthusiastic restoration work. Generally the more pentimenti, the more likely a painting is to be the prime version. It also gives useful insights into working practices.

Among many other changes in the Arnolfini Portrait of 1434 (right), his face was higher by about the height of his eye, hers was higher, and her eyes looked more to the front. Each of his feet was underdrawn in one position, painted in another, and then overpainted in a third. These alterations are seen in infra-red reflectograms.

Similar uses of infrared are made by historians on various types of objects, especially very old written documents such as the Dead Sea Scrolls, the Roman works in the Villa of the Papyri, and the Silk Road texts found in the Dunhuang Caves. Carbon black used in ink can show up extremely well.

Biological Systems

The pitviper has a pair of infrared sensory pits on its head. There is uncertainty regarding the exact thermal sensitivity of this biological infrared detection system.

Other organisms that have thermoreceptive organs are pythons (family Pythonidae), some boas (family Boidae), the Common Vampire Bat (*Desmodus rotundus*), a variety of jewel beetles (*Melanophila acuminata*), darkly pigmented butterflies (*Pachliopta aristolochiae* and *Troides rhadamantus plateni*), and possibly blood-sucking bugs (*Triatoma infestans*).

Photobiomodulation

Near infrared light is currently used for treatment of chemotherapy induced oral ulceration as well as wound healing. There is some work relating to anti herpes virus treatment. Research projects include work on central nervous system healing effects via cytochrome c oxidase upregulation and other possible mechanisms.

HEALTH HAZARD

Strong infrared radiation in certain industry high heat settings may constitute a health hazard to the eyes and the vision. More so, since the radiation is invisible. Therefore special IR proof protective eyeglasses must be worn in such places.

The Earth as an Infrared Emitter

The Earth's surface and the clouds absorb visible and invisible radiation from the sun and re-emit much of the energy as infrared back to the atmosphere. Certain substances in the atmosphere, chiefly cloud droplets and water vapour, but also carbon dioxide, methane, nitrous oxide, sulfur hexafluoride, and chlorofluorocarbons, absorb this infrared, and re-radiate it in all directions including back to Earth. Thus the greenhouse effect keeps the atmosphere and surface much warmer than if the infrared absorbers were absent from the atmosphere.

Chapter 12

Far-Infrared Spectra of Lateral Quantum Dot Molecules

Far-infrared (FIR) magneto-optical absorption spectroscopy is one experimental technique to study electrons confined in semiconductor quantum dots (QDs)[1, 2]. It was, however, realized at the early stage of QD research that FIR spectroscopy is unable to reveal interesting many-electron effects in parabolic-confinement QDs. This is because the electromagnetic waves couple only to the centre-of-mass variables of electrons. The resulting FIR spectrum is rather simple, showing two branches, $\omega_{\pm}$, as a function of magnetic field.

These branches, one with positive energy dispersion (ω_+) and one with negative energy dispersion (ω_-) as a function of magnetic field, are called the Kohn modes. The spectrum does not depend on the number of electrons nor on the interactions between them. This condition in parabolic QDs is called the generalized Kohn theorem.

The condition can be lifted if the confinement is not parabolic, and many experiments show more complex FIR spectra , -. Also calculations - have shown non-trivial FIR spectra of non-parabolic QDs. It is also possible that spin-orbit interaction and impurities near QDs can have an effect on the FIR spectrum.

In a non-parabolic QD, the relative internal motion of electrons could be accessible with the FIR spectroscopy, but recent studies suggest that additional features in the FIR spectra are still of collective nature. The interpretation of the observed FIR spectra of a non-parabolic QD is usually far from trivial. It is clear that deviations arise from non-parabolic confinement, but the detailed cause of the deviations, thus the interpretation of spectrum, is not always straightforward. It is especially interesting to see how many-electron interactions appear in the FIR spectra of QDs.

In the calculation of FIR spectra of two-electron lateral double QD and lateral four-minima QD molecule (QDM) it was shown that the deviations from the Kohn modes arise mainly from non-parabolic confinement and interactions have only minor effects on the spectra. Effects of relative motion of electrons on FIR spectra were studied by turning electron-electron interactions on and off.

The results indicate that the FIR spectra of QDMs reveal mainly the centre-of-mass collective excitations of electrons. Actually more pronounced deviations from Kohn modes are observed when interactions between the electrons are turned off.

This is an interesting result, suggesting that fewer features are observed with more electrons in a non-parabolic QD. These studies, however, include only two electrons and the FIR spectra may depend on particle number. Also as more single-particle levels are occupied the centre-of-mass excitations with slightly different energies may be observed where interactions do not, necessarily, have any significant role. Calculations where electron interactions can be turned on and off could show how the electron-electron interactions show up in FIR spectra for greater electron numbers ($N > 2$).

In this paper, we study three different lateral QDM confinements. We study two-minima QDM (double dot), and square-symmetric and rectangular-symmetric four-minima QDMs. We calculate the FIR spectra for all the QDM confinements and compare them to the parabolic-confinement QD FIR spectrum.

We analyse in detail the effect of electron-electron interactions on the FIR spectra. We use a very accurate exact diagonalization technique for interacting electrons but limit our studies to two interacting electrons. We also calculate the

FIR spectra of non-interacting electrons up to six electrons by occupying single-particle levels and calculating dipole-transition elements between the single-particle levels.

These provide some insight on how FIR spectra are modified with greater electron numbers, even if we do not include electron-electron interactions.

In particular if excitations are collective also with the greater electron numbers, the FIR spectra are basically given by the transitions between single-particle levels of the QD confinement in question.

Model and Method

We model the interacting two-electron QDM with a two-dimensional Hamiltonian

$$H = \sum_{i=1}^{2} \left[\frac{\left[-i\hbar\nabla_i - (e/c)\mathrm{A}\right]^2}{2m^*} + V_c(r_i) \right] + \frac{e^2}{\in r_{12}},$$

where V_c is the external confinement potential of the QDM taken to be

$$V_c(r) = \frac{1}{2} m^* \omega_0^2 \min\left\{ \sum_j^M (r - L_i)^2 \right\}.$$

The coordinates are in two dimensions r = (x, y), the L_j's ($L_j = (L_x, L_y)$) give the positions of the minima of QDM potential and M is the number of minima. When $L_1 = (0, 0)$ (and $M = 1$) we have a single parabolic QD. With $M = 2$ and $L_{1,2} = (\pm L_x, 0)$ we get a double-dot potential. We also study four-minima QDM ($M = 4$) with minima at four possibilities of $(\pm L_x, \pm L_y)$ (see figure 1). We study both square-symmetric ($L_x = L_y$) and rectangular-symmetric ($L_x \neq L_y$) four-minima QDMs. The confinement potential can also be written using the absolute values of x and y coordinates as

$$V_c(x,y) = \frac{1}{2} m^* \omega_0^2 \times \left(r^2 - 2L_x[x] - 2L_y|y| + L_x^2 + L_y^2\right)$$

For nonzero L_x and L_y, the perturbation to the parabolic potential comes from the linear terms of L_x or L_y containing also the absolute value of the associated coordinate.

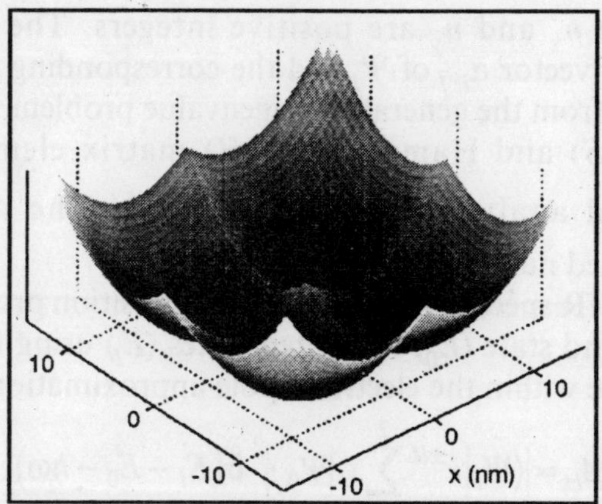

Fig.Confinement Potential of Square-symmetric ($L_x = L_y$ = 5nm) four-minima QDM.

We use the GaAs material parameters $m^*/m_e = 0.067$ and $\hbar$ ε = 12.4, and confinement strength $h\omega_0$ = 3.0meV. This confinement corresponds to harmonic oscillator strength of $l_o = \sqrt{\hbar/\omega_o m^*} \approx 20$ nm. We concentrate on closely coupled QDMs where $L_{x,y} \le l_0$. The magnetic field (in the *z*-direction) is included in the symmetric gauge by vector potential A. The Hamiltonian of equation (1) is spin-free, and the Zeeman energy can be included in the total energy afterwards ($E_Z = g^*\mu_B B S_Z$ with g^* = -0.44 for GaAs). We disregard the three-fold splitting of each triplet state (S_Z = 0, ±1) and consider only the lowest energy one (S_Z = 1).

We exclude the explicit spin-part of the wavefunction and expand the many-body wavefunction in symmetric functions for the spin-singlet state (S = 0) and anti-symmetric functions for the spin-triplet state (S = 1)

$$\Psi_s(\mathrm{r}_1,\mathrm{r}_2) - \sum_{i \leqslant j} \alpha_{i,j}\left[\phi_i(\mathrm{r}_1)\phi_j(\mathrm{r}_2) + (-1)^s \phi_j(\mathrm{r}_i)\right],$$

where $\alpha_{i,j}$s are complex coefficients. The one-body basis functions $\phi_i(\mathrm{r})$ are two-dimensional Gaussians

$$\phi_{n_x,n_y}(\mathrm{r}) = x^{n_x} y^{n_y} e^{-r^2/2},$$

where n_x and n_y are positive integers. The complex coefficient vector $\alpha_{i,j}$ of Ψ_l and the corresponding energy E_l are found from the generalized eigenvalue problem where the overlap (S) and Hamiltonian (H) matrix elements are calculated analytically $(H\alpha_1 = E_1 S\alpha_1)$. The matrix is diagonalized numerically.

The FIR spectra are calculated as transition probabilities from ground state (E_0) to excited states (E_l) using the Fermi golden rule within the electric-dipole approximation

$$A_{l\pm} \propto \left|\left\langle \Psi_l \right| e^{\pm i\phi} \sum_{i=1}^{2} \mathrm{r}_i \left| \Psi_0 \right\rangle\right|^2 \delta\left(E_l - E_0 - \hbar\omega\right).$$

We assume circular polarization of the electromagnetic field: $e^{\pm i\phi}\Sigma_i r_i = \Sigma_i(x_i \pm iy_i) = z_{\pm}$, where plus indicates right-handed polarization and minus left-handed polarization. The results are presented for non-polarized light as an average of the two circular polarizations.

FIR spectra of two-electron QDMs

The dipole-allowed magneto-optical excitation spectrum of an isolated harmonic-confined QD consist of two branches ω_+ and ω_-, whose energy dispersion is well understood and does not depend on the number of electrons in the QD:

$$\Delta E_{\pm} = \hbar\omega_{\pm} = \hbar\sqrt{\omega_0^2 + \left(\omega_c/2\right)^2} \pm \hbar\omega_c/2,$$

where ω_0 describes the external confinement, $\omega_c = eB/m^*$ is the cyclotron frequency and m^* is the effective mass of the electron. Fock-Darwin energy levels as a function of magnetic field are given by $E_{nl} = \left(2n + |l| + 1\right)\hbar\omega = \frac{1}{2} l\hbar\omega_c$, where $n = 0, 1, 2, \ldots$ and $l = 0, \pm1, \ldots$ are principal and azimuthal quantum numbers, respectively, and $\omega = \sqrt{\omega_0^2 + \left(\omega_c/2\right)^2}$. In the dipole-allowed transitions between Fock-Darwin states angular momentum must change by unity, $\Delta l = \pm1$. figure2 shows relative transition probabilities between Fock-Darwin energy levels $P = \left\langle \phi_{n',l\pm1} \right| e^{\pm i\phi} r \left| \phi_{n,l} \right\rangle$.

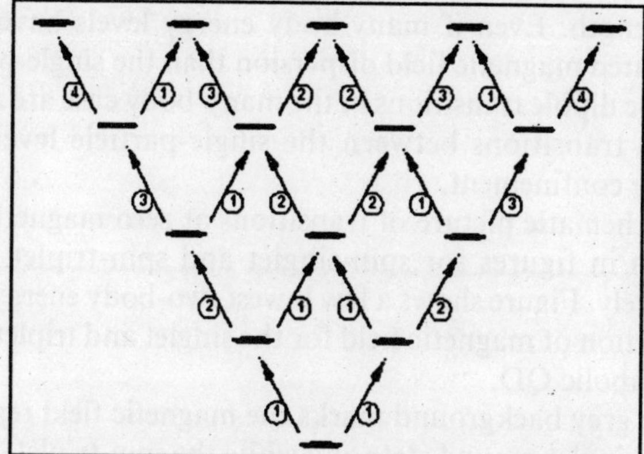

Fig.Relative Transition Probabilities between

Fock-Darwin Energy Levels $P = \left\langle \phi_{n',l\pm1} \middle| e^{\pm i\phi} r \middle| \phi_{n,l} \right\rangle$.

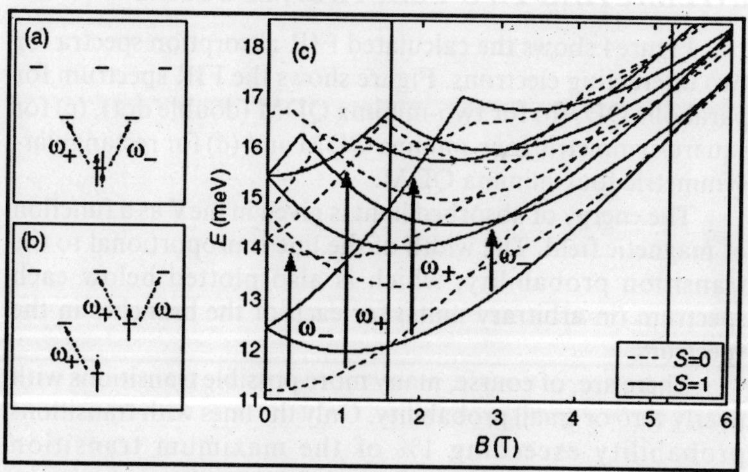

Fig.Transitions for spin singlet (a) and spin triplet (b). (c) Two-particle energy levels of a parabolic QD. The red lines show singlet energy levels and blue lines triplet energy levels. Grey background shows singlet ground-state regions and white triplet ground-state regions. Dipole transitions from one spin type to another are forbidden.

For a two-electron QD, the ground state can be either spin singlet ($S = 0$) or spin triplet ($S = 1$) depending on the magnetic field strength. Even if many-body energy levels have more complicated magnetic field dispersion than the single-particle levels, the dipole transitions in the many-body case are always equal to transitions between the single-particle levels in a *parabolic* confinement.

A schematic picture of transitions at zero magnetic field is shown in figures for spin-singlet and spin-triplet states, respectively. Figure shows a few lowest two-body energy levels as a function of magnetic field for the singlet and triplet states of a parabolic QD.

The grey background marks the magnetic field region of the spin-singlet ground state and white the spin-triplet ground state. Dipole transitions from one spin type to another are forbidden.

INTERACTING TWO-ELECTRON SPECTRA

Figure4 shows the calculated FIR absorption spectra for two interacting electrons. Figure shows the FIR spectrum for parabolic QD, (b) for two-minima QDM (double dot), (c) for square-symmetric four-minima QDM and (d) for rectangular-symmetric four-minima QDM.

The energy of absorbed light is given in meV as a function of magnetic field. The width of the line is proportional to the transition probability, which is also plotted below each spectrum (in arbitrary units) for each of the branches in the spectrum.

There are, of course, many more possible transitions with nearly zero or small probability. Only the lines with transitions probability exceeding 1% of the maximum transition probability are included in figure4. Large open circles in the spectra represent the two Kohn modes, ω_+ and ω_-, plotted with 1T spacings.

In QDMs there are ground-state transitions from spin-singlet ($S = 0$) to spin-triplet ($S = 1$) ground states, and the other way round, as a function of magnetic field[19, 20]. Vertical lines indicate the singlet-triplet (or triplet-singlet) transition points and the singlet ground-state regions of spectrum are marked with red colour and the blue lines denote triplet regions. Dipole transitions from one spin type to another are forbidden.

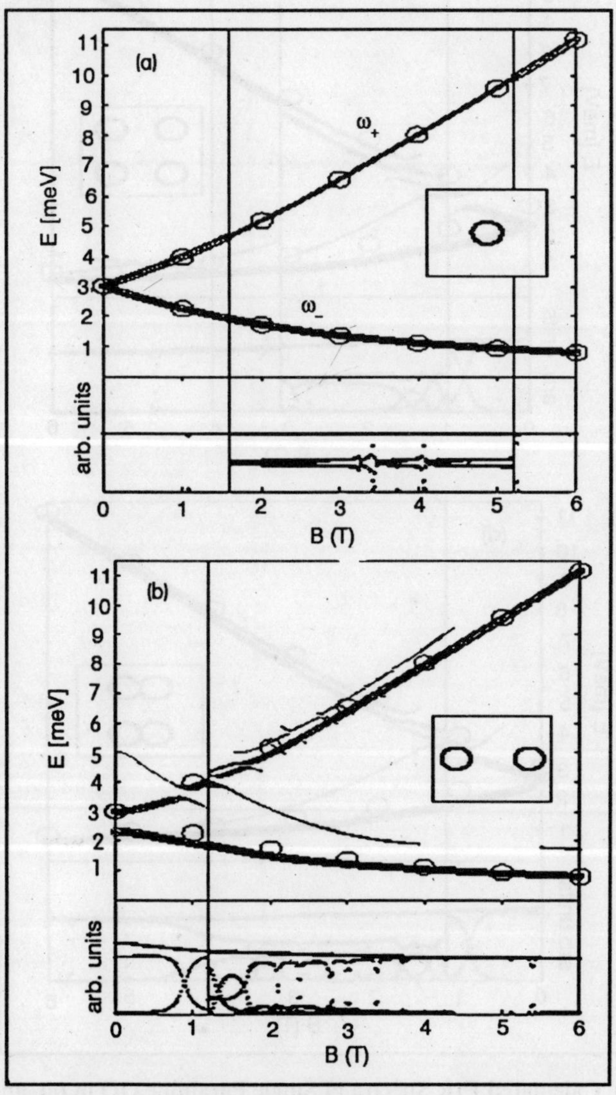

Fig. Calculated FIR spectra of single parabolic QD in (a), double dot with L_x = 15nm in (b), square-symmetric four-minima QDM with $L_x = L_y$ = 10nm in

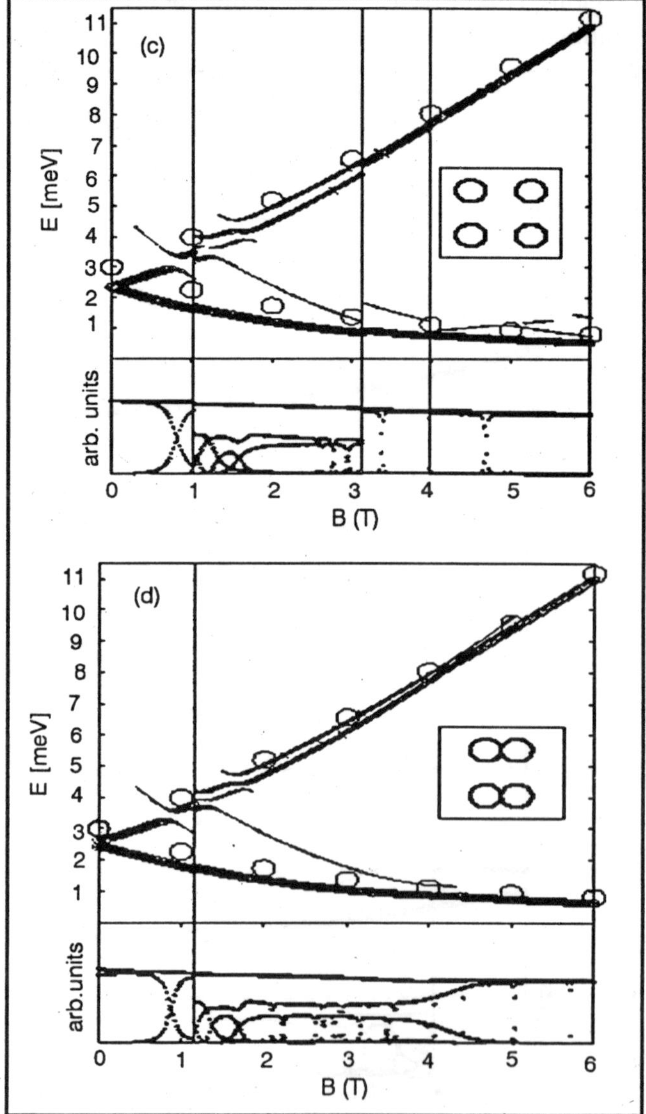

Fig. Calculated FIR Spectra of Single Parabolic QD in (c), and Rectangular-symmetric Four-minima QDM with $L_x = 5$, $L_y = 10\text{nm}$ in (d). The Energy of Absorbed Light is Given in meV as a Function of Magnetic Field.

The width of the line is proportional to the transition probability, which is also plotted below each spectrum (in arbitrary units) for each of the branches in the spectrum. Transitions with small probability are excluded from the spectra. Only the lines with transitions probability exceeding 1% of the maximum transition probability are included in the spectra. Large open circles in the spectra represent two Kohn modes plotted with 1T spacings. Vertical lines indicate the singlet-triplet (or triplet-singlet) transition points.

Figure 4(a) shows the calculated FIR spectra for a single parabolic QD. The dipole transition probabilities are calculated from the two-body ground-state level to higher two-body energy levels as shown in figure3(c). The spectrum of figure4(a) shows two Kohn modes ω_+ and ω_-. These coincide perfectly with the open circles presenting Kohn modes, as they should, since in a parabolic QD the FIR spectra does not depend on the number of electrons in the QD nor on the interactions between them. The magnitude of the absorbed light can only change when the number of electrons in the QD is changed[1, 4]. However, in the two-electron QD of figure4(a) the width of the upper branch is halved after the first singlet triplet transition. The transition line is split into two degenerate transitions with almost equal transition probabilities, which, if summed up, would equal to the transition probability of ω_- . In the calculations we can identify two degenerate levels to which electrons can be excited (see figure3(b)), but in experiments the observed quantity is the overall absorption of the energy in question. Therefore, one can observe Kohn modes with constant transition probabilities as a function of magnetic field.

figures4(b)-(d) show FIR spectra for QDMs. Now as the symmetry of the confinement is lower, the centre of mass and relative motion do not decouple. There are clear deviations from the Kohn modes. In all of the QDM potentials we can see some general features. The main branches of the spectra always lie below the Kohn modes (below open circles). There are clear anti-crossings and the upper branch is split to two in some parts of the triplet spectra.

In the double dot spectra of figure4(b), the excitations at zero magnetic field are identified to a lower energy one along the long axis (the line connecting two dots, x-axis) and a higher energy one along the short axis (y-axis). This can be also verified by calculating the absorption of linearly polarized light. At $B = 0$ the excitation of ω_+ is unaffected by the interactions between electrons and coincides with the confinement ($\hbar\omega_0 = 3$meV), because in the double dot the potential is still parabolic along the y-axis. However, with nonzero B this is no longer true since a magnetic field mixes the two linear polarizations. At high magnetic field the spectrum approaches Kohn modes and no anti-crossings are observed. In the high field region the electrons are effectively more localized into individual dots. Then the non-parabolic nature of the confinement potential has a less important role, since the electron density is more localized close to the parabolic minima.

After the singlet-triplet transition there appears an additional level, ω_{+2}, above ω_+ with a lower transition probability. A simple explanation for two positive dispersion levels would be in the picture of figure3(b) where the two transitions of ω_+ mode could have slightly different energy difference in a non-parabolic confinement. However, the interpretation is not so straightforward, since the dipole transitions are calculated between two-body levels and not occupied single-particle levels, as we will see when we analyse interactions in the next subsection. Another deviation from the Kohn modes of a parabolic QD are seen as small discontinuities in the energy of the spectral lines when the singlet changes to triplet. However, this discontinuity is very small and may not be resolved in experiments. See for more details on double dot FIR spectra.

In the square-symmetric four-minima QDM (figure4(c)) the potential is identical in the x- and y-directions resulting in only one excitation at $B = 0$ as the long and short axes are equal. This excitation is clearly lower than the $\hbar\omega_0 = 3$meV confinement energy of a single minimum. The upper branch is split to two levels in the first triplet region at magnetic field values between 1 and 3T, but in the second triplet region $B >$ 4T there is only one upper branch. In square-symmetric four-minima QDM the two levels of the split-off upper branch have almost equal transition probabilities. Discontinuities in the spectra appear in the singlet-triplet (or triplet-singlet) transition points. A brief representation of the FIR spectra of square-symmetric four-minima QDM can also be found in.

The spectrum of rectangular-symmetric four-minima QDM, in figure4(d), has similar features as the double dot spectra. There is a gap between the two branches at $B = 0$. The lower branch at $B = 0$ in figure4(d) corresponds to excitation along the y-axis and the upper branch along the x-axis.

The upper mode does not start at 3meV, as in the double dot, since the potential is not parabolic in the x-direction. After the singlet-triplet transition an additional mode, ω_{+2}, appears above ω_+. The transition probability of ω_{+2} is higher compared to the ω_{+2} of double dot, but the transition probability is not as high as in the square-symmetric four-minima QDM. The modes of figure4(d) at high magnetic field do not become so close to Kohn modes as in the double dot. The reason is that electrons localize into two decoupled double dots rather than into single QDs as in the double dot spectrum of figure4(b).

All QDM confinements show anti-crossings in ω_+. The clearest anti-crossings are seen at low magnetic field strengths. At higher B there are still anti-crossings but the energy gap is so small that they are not visible in the spectra. The transition probabilities in the lower panels, however, reveal anti-crossings as the transition probability of one mode decreases and the other one increases as a function of magnetic field. The biggest anti-crossing gap in energy is seen with square-symmetric four-minima QDM. This is interesting, since in our previous studies the square-symmetric four-minima QDM most resembled a parabolic QD, whereas double dot and rectangular four-minima QDM showed greater deviations in the ground-state properties and in the low-lying eigenstates. Obviously this is not true with the excitation spectrum and with the higher eigenstates.

Analysis of Interactions in Two-Electron FIR Spectra

We conclude that different types of deviations appear in the spectrum when the confinement potential is not perfectly parabolic. In order to analyse these deviations in more detail, we plot the full spectra up to 6T magnetic field for both singlet and triplet states with two interacting electrons and also for two non-interacting electrons for the comparison. We are interested to see which features in the spectra are due to electron-electron interactions and which result from the lower symmetry of the confinement potential.

It is also interesting to see if some features in the spectra can be used to identify the confinement potential. This would be useful for experiments where the profile of the underlying confinement potential is not clear.

We will first analyse the two-electron singlet spectra shown in figures 5(a)-(c) for interacting electrons and in (d)-(f) for non-interacting electrons. The two-electron triplet spectra are shown in figures6(a)-(c) and the corresponding non-interacting spectra in (d)-(f).

In a real two-electron QDM system, singlet-triplet transitions must be taken into account and therefore experimentally observable spectra for two interacting electrons are shown in figure4. As the non-interacting singlet spectra, in figures5(d)-(f), are the same as the single-particle spectra, these would correspond to experimental FIR spectra with one electron in the QDM.

The non-interacting triplet, on the other hand, corresponds to the FIR spectra of two occupied single-particle levels. The double dot singlet spectrum in figure 5(a) looks qualitatively very similar to the non-interacting spectrum in figure 5(d). The first difference is that the Kohn modes lie at the lower energy in the non-interacting spectra compared to the interacting spectra. The only exception is in the upper mode at $B = 0$ where both have excitation energy of $\hbar\omega_0 = 3$meV. In zero magnetic field this corresponds to linearly polarized excitation along the y-axis where the potential is parabolic and therefore excitations are not affected by the electron-electron interactions.

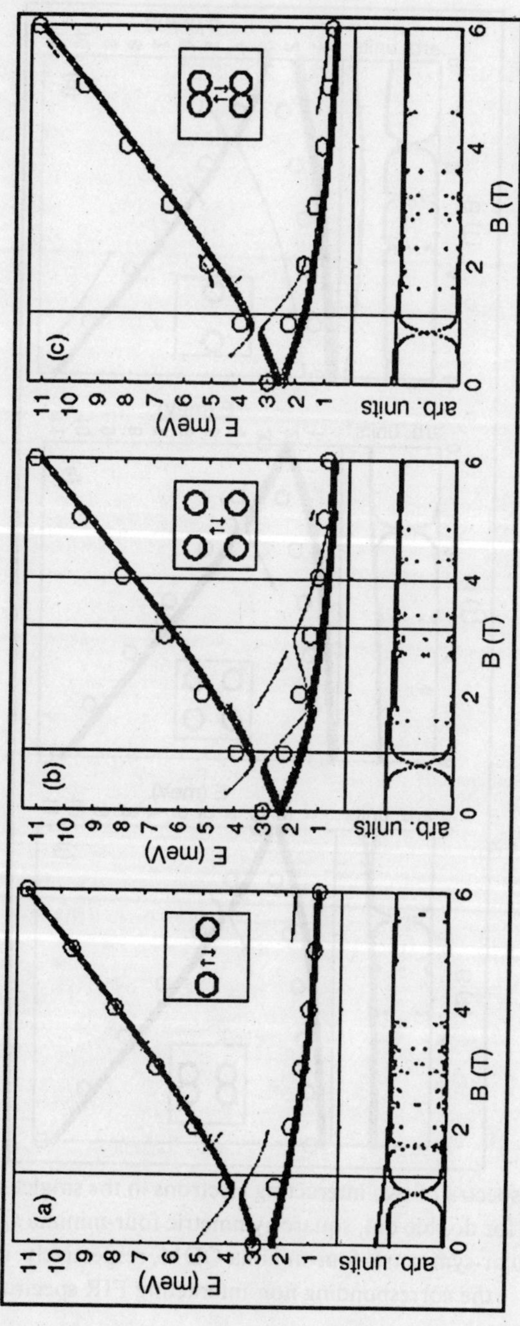
(a)
(b)
(c)
E (meV)
arb. units
B (T)

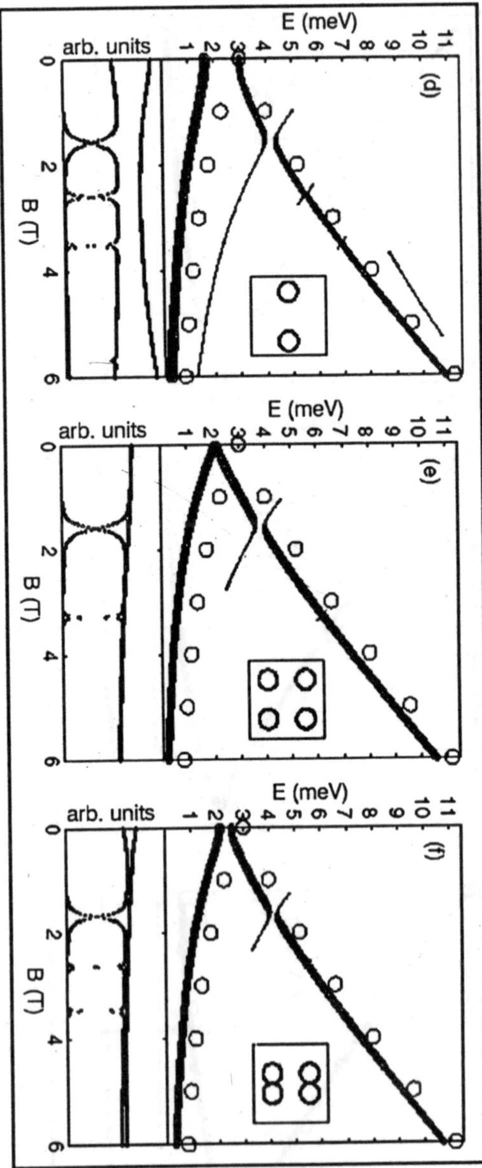

Fig. FIR spectra of two interacting electrons in the singlet symmetry in (a)-(c) for double dot, square-symmetric four-minima QDM, and rectangular-symmetric four-minima QDM, respectively. (d)-(f) show the corresponding non-interacting FIR spectra.

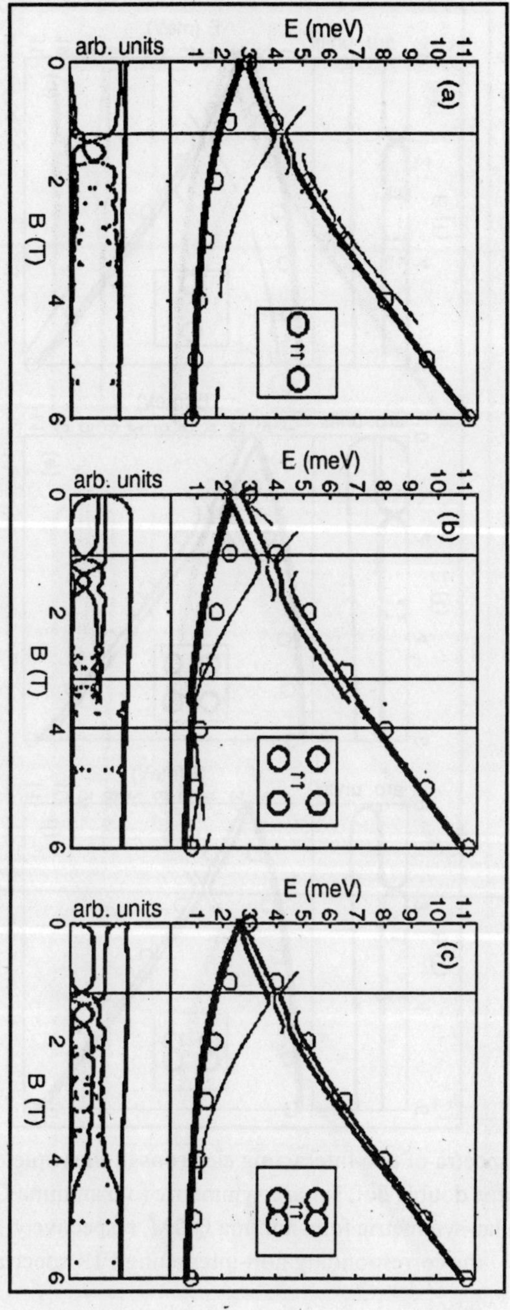

(a)
(b)
(c)
E (meV)
arb. units
B (T)

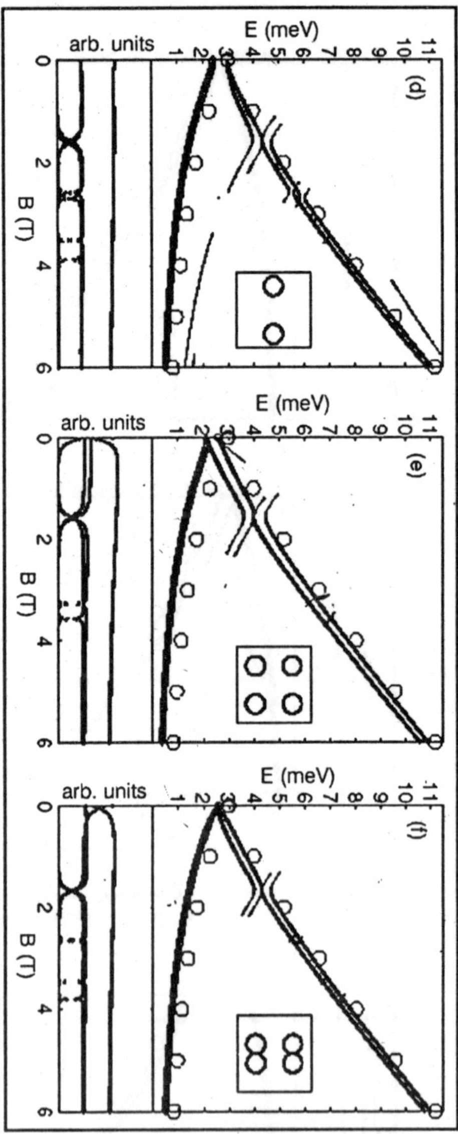

Fig. FIR spectra of two interacting electrons in the triplet symmetry in (a)-(c) for double dot, square-symmetric four-minima QDM, and rectangular-symmetric four-minima QDM, respectively. (d)-(f) show the corresponding non-interacting FIR spectra.

Another difference is the zero-field gap between Kohn modes which is clearly greater in the non-interacting case. Higher energy excitations in the interacting case can be explained with the Coulomb repulsion between the electrons. Coulomb repulsion effectively steepens the confinement resulting in higher excitation energies.

The anti-crossing points are also affected by interactions whereas the gap remains almost the same. The anti-crossings can be seen at lower magnetic field values in the interacting spectrum. Qualitatively the two spectra look very similar and we conclude that the deviations observed in the spectrum result from the low symmetry confinement and interactions only shift the excitation energies and change the anti-crossing points.

Generally the same conclusions drawn for the double dot hold for the square-symmetric four-minima QDM singlet spectra in figures 5(b) and (e). Now in the square-symmetric four-minima QDM the zero-field excitation is degenerate as the two perpendicular directions (x and y) have identical confinement profiles. Interactions shift the zero-field excitation to higher energy due to Coulomb repulsion. The first anti-crossing is again seen at lower B in the interacting spectrum. However, this time the anti-crossing gap is somewhat greater in the interacting case. As a new feature, compared to the double dot, there are now discontinuities in the interacting spectrum. The discontinuities in (b) are at the transition points when two singlet states with a different symmetry cross. At $B \approx 1.8$T the lowest energy singlet state changes from no-vortex state to two-vortex state and the discontinuity at $B \approx 5.2$T is at the crossing point of two-vortex and four-vortex states (see for more details on vortex states). However, the discontinuities are seen at the magnetic field values where the true ground state is triplet and therefore the discontinuities in figure 5(b) would not be observed in FIR experiments.

In the rectangular-symmetric four-minima QDM the confinement profile is non-identical and non-parabolic in both perpendicular directions. Therefore both zero-field excitations are lower than 3meV and there is also a gap between the two modes. Again excitation energies as a function of magnetic field are higher in the interacting spectrum and anti-crossings are seen at the lower magnetic field values compared to the non-interacting spectrum. Interesting point, when comparing interacting and non-interacting spectra, is that the zero-field excitation energy of the lower mode shifts to 0.30 meV higher and the higher mode shifts only 0.09 meV higher when the interactions are turned on. It means that the Coulomb repulsion is over three times more significant in the excitation along the long axis (y-axis) than it is along the short axis (x-axis). Electrons are localized into two distant double dots. Coulomb repulsion is more important between the two double dots than it is inside one double dot, i.e., along long and short axes.

Next, we do similar comparisons for the triplet spectra of QDMs. The interacting spectra are shown in figure 6(a)-(c) and non-interacting in figure 6(d)-(f). We will first compare double dot interacting triplet spectra of figure 6(a) to the non-interacting spectra of figure6(d). In general, the same analysis applies to triplet spectra as for the singlet spectra of double dot. However, in the triplet state there is an additional mode, ω_{+2}, above the main branch, ω_+, in figure 6(a). The additional mode has clearly weaker transition probability than the ω_+ mode. In the non-interacting case the upper mode is split to two modes which both have equal transition probabilities. In the interacting spectrum the upper mode has lower transition probability and soon after 4T it completely vanishes from the spectrum.

In the square-symmetric four-minima QDM also the interacting triplet spectrum, as well as the non-interacting spectrum, has two upper modes with almost equal transition probabilities. In four-minima QDM the split-off branch suddenly changes to one mode at $B \approx 3.5$T. There the triplet ground state changes from the one-vortex to three-vortex solution (see for more details on vortex states). Now the two triplet states actually cross and there occurs a discontinuity in the FIR spectrum. Yet as the singlet is the true ground state of the square-symmetric four-minima QDM between B 3 and 4T, the discontinuity of figure 6(b) cannot be seen in experiments. In triplet spectra there are much smaller differences between the excitation energies of interacting and non-interacting electrons compared to the singlet spectra. In the singlet state electrons of opposite spins can occupy the same single-particle levels in the many-body configurations and therefore Coulomb repulsion is more significant for the singlet state. In other words, the Pauli exclusion principle keeps the same-spin electrons further apart and therefore Coulomb repulsion has a lesser effect on the triplet state compared to the singlet state.

In the rectangular-symmetric four-minima QDM, the zero-field gap is really small and not visible in the interacting spectra of figure 6(c) and neither in the non-interacting triplet spectra of figure 6(f). The transition probabilities of the two main branches (ω_+ and ω_-), in the interacting spectrum, are almost equal at zero field, but soon after the magnetic field strength increases the transition probability of ω_+ decreases at the same time as the transition probability of ω_{+2} increases. After B = 4T the ω_{+2} starts to weaken again as the symmetry of the triplet ground state is changing. The ω_{+2} dies out continuously. The two upper modes of the non-interacting spectrum have almost equal transition probabilities. The interacting modes are higher in energy, also at $B = 0$, and anti-crossings shift to lower B in the interacting spectrum. Anti-crossing gaps are not noticeably affected by the interactions.

FIR SPECTRA OF NON-INTERACTING ELECTRONS

A general conclusion drawn above is that the non-interacting spectra show more deviations from the Kohn modes than the interacting spectra. All the same features are present in non-interacting and interacting spectra. In addition, the non-interacting spectra show in all cases similar structure as the interacting spectra. Interactions only shift some of the features seen in the spectra. Therefore we calculate FIR spectra of non-interacting electrons up to $N = 6$ electrons in the QDMs to see if more electrons, even if non-interacting, produce changes in the FIR spectra.

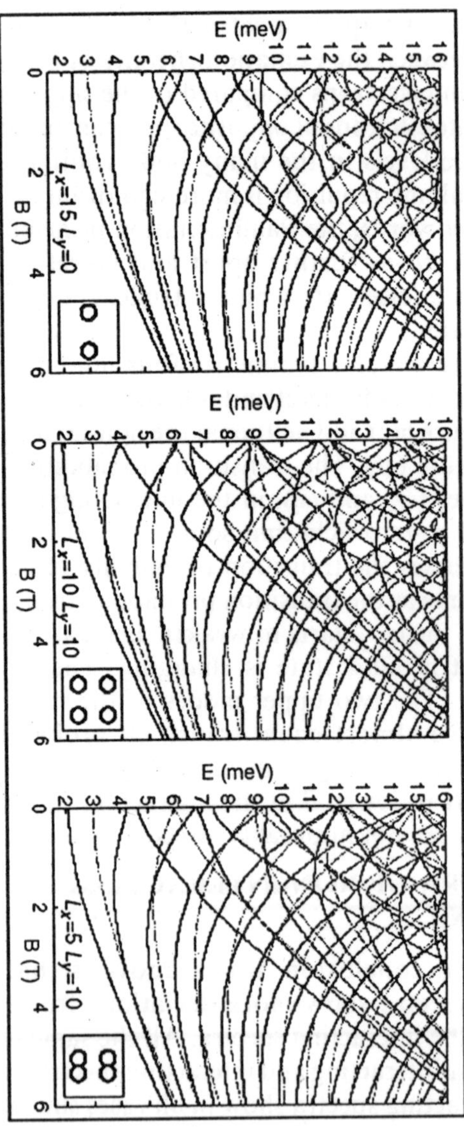

Fig. Single-particle energy levels of double dot (L_x = 15, L_y = 0nm), four-minima square-symmetric QDM (L_x = L_y = 10nm) and four-minima rectangular symmetric QDM (L_x = 5, L_y = 10nm). Dotted lines show single-particle (Fock-Darwin) energy levels of parabolic ($\hbar\,\omega_0$ = 3meV) QD.

Before presenting the FIR spectra we plot single-particle energy levels of all three QDM confinements in figure 7. The solid lines show QDM energy levels and the dotted lines Fock-Darwin energy levels of a parabolic QD. In QDMs the energy levels shift to lower energies compared to parabolic QD. Also many anti-crossings and zero-field splittings of energy levels are visible in QDMs.

We obtain non-interacting FIR spectra of QDMs by occupying lowest single-particle energy levels with N non-interacting electrons and calculating dipole transitions to higher unoccupied levels.

The non-interacting FIR spectra are shown in figures 8,9 and10 for a double dot, square-symmetric four-minima QDM and rectangular-symmetric four-minima QDM, respectively. (a)-(d) correspond to FIR spectra of $N = 1$-6 electrons confined in a QDM. $N = 1$ and $N = 2$ are also shown in figures 5 and 6. The transition probabilities are multiplied when more electrons are added in the QDM. We divide each transition probability with N in order to ease comparisons.

We will first analyse the single-particle spectra of a double dot. figure8(c) for $N = 3$ shows an additional mode below ω_- in the low-field region with the excitation energy below $E =$ 2meV. Also another mode is visible with a small transition probability which has zero excitation energy at B 0.9T.

If we study the single-particle energy levels of figure7(a), one can see that with the three lowest levels occupied, the uppermost levels cross at $B \approx 0.9$T. When these levels cross, the weak mode has excitation energy of $\Delta E = 0$ and the transition probability with the E d" 2meV mode vanishes.

Another feature that is observed with $N = 3$, but absent in two-particle spectra, is an anti-crossing in ω_-. This anti-crossing is opposite to the anti-crossings in ω_+ where the low-field mode curves upwards and the high-field mode curves from down to up while increasing its strength. ω_+ with three electrons looks quite similar to ω_+ in $N = 2$ spectrum. In $N =$ 4 spectrum of figure 8(d) there is one level below ω_- at low B but the clear anti-crossing, present in $N = 3$, is missing from the spectrum. After $B \approx 1.7$T there is just one lower mode.

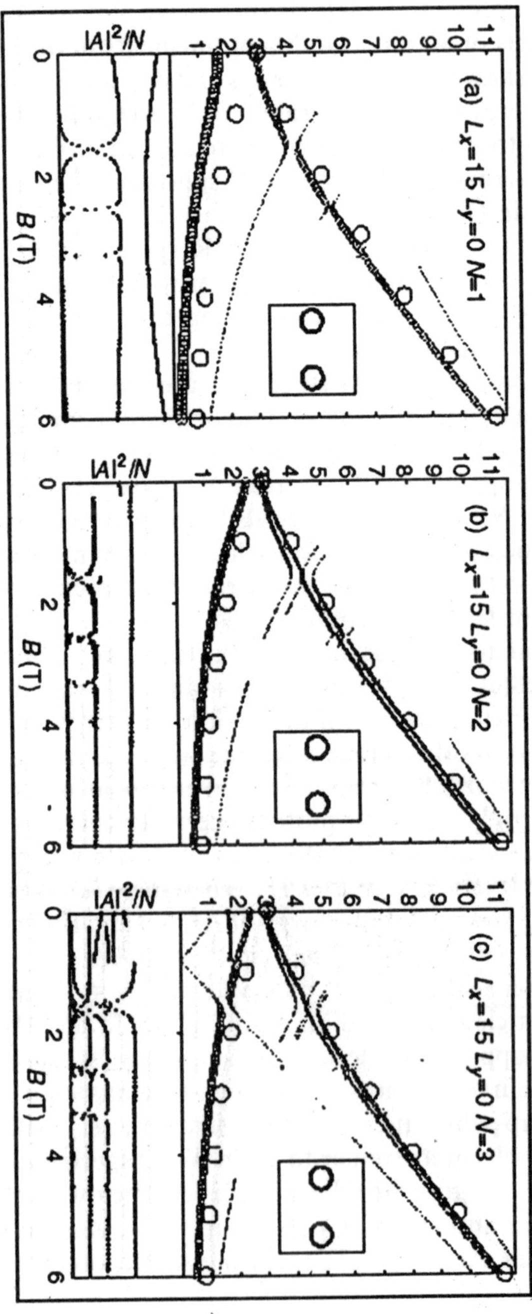
(a) Lx=15 Ly=0 N=1
|A|²/N
B (T)
(b) Lx=15 Ly=0 N=2
|A|²/N
B (T)
(c) Lx=15 Ly=0 N=3
|A|²/N
B (T)

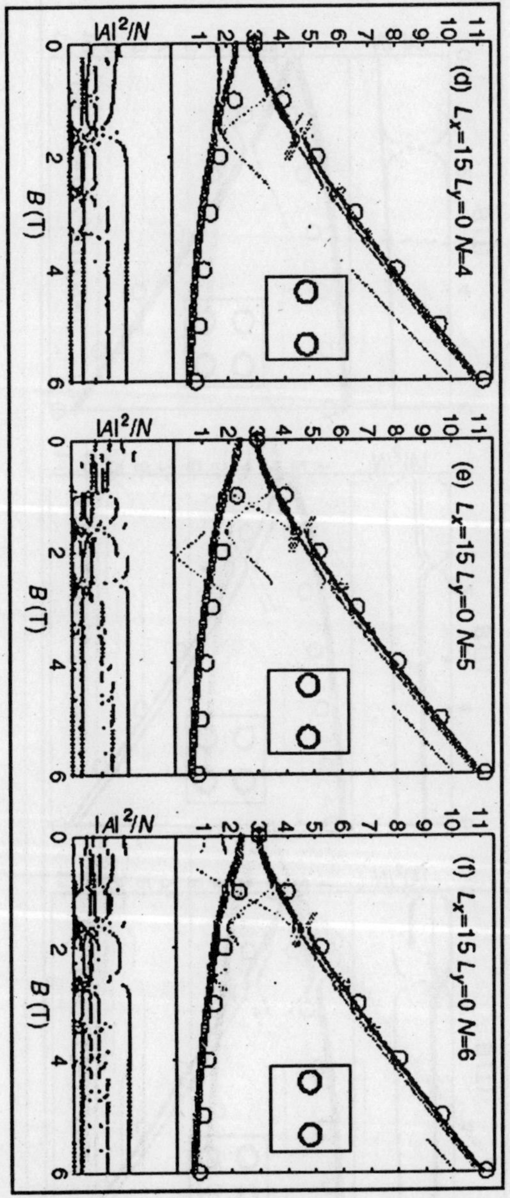

Fig.FIR Spectra for N = 1-6 Non-interacting Electrons in Double Dot (L_x = 15, L_y = 0nm).

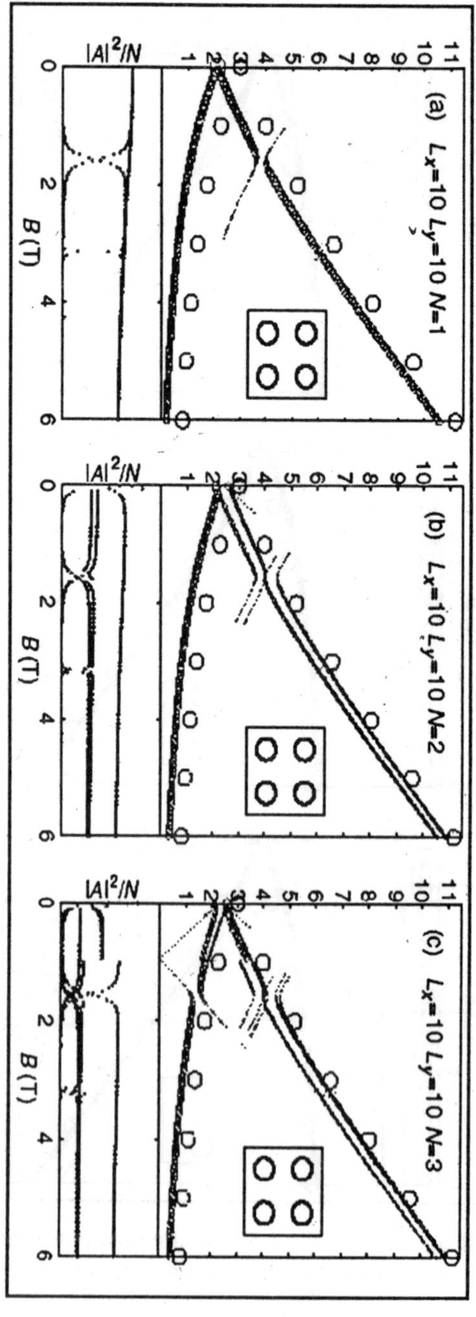

(a) Lx=10 Ly=10 N=1
(b) Lx=10 Ly=10 N=2
(c) Lx=10 Ly=10 N=3
|A|²/N
B (T)

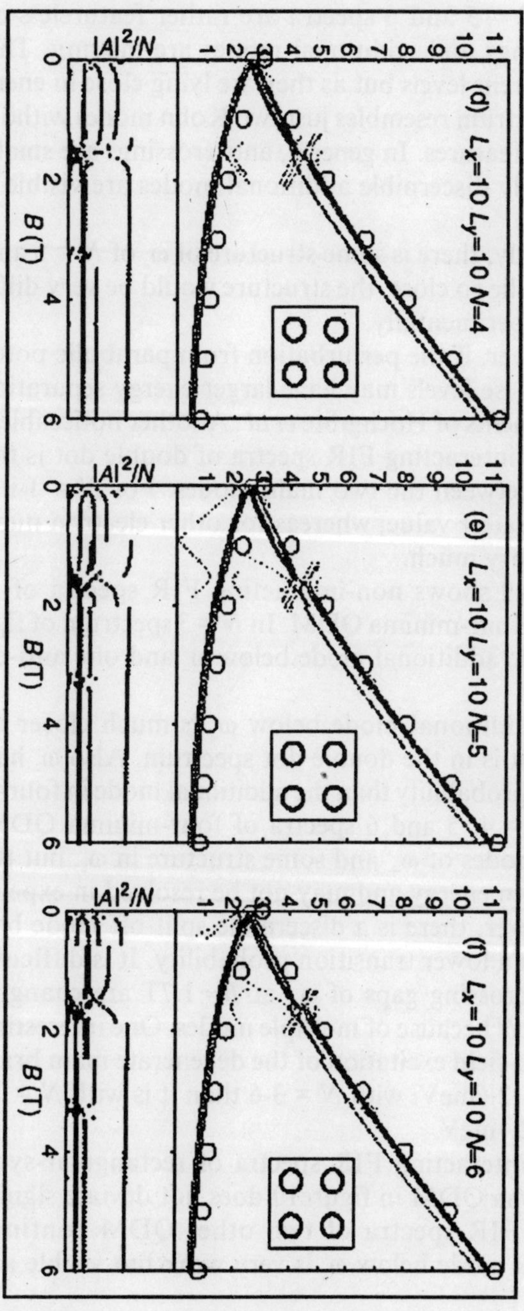
(d) Lx=10 Ly=10 N=4
(e) Lx=10 Ly=10 N=5
(f) Lx=10 Ly=10 N=6
|A|²/N
B(T)

The N = 5 and 6 spectra are rather featureless as clear anti-crossings and additional modes are missing. There are many different levels but as they are lying close in energy, the overall spectrum resembles just two Kohn modes without clear additional features. In general, anti-crossings are smaller and fewer clearly discernible additional modes are visible with N = 5 and 6.

Actually, there is some structure of ω_- of N = 5 and 6 but as energies lie so close, the structure would be very difficult to observe experimentally.

However, if the perturbation from parabolic potential is stronger these levels may have larger energy separation as in the experiments of Hochgrδfe *et al*. Another noticeable feature in the non-interacting FIR spectra of double dot is the zero-field gap between the two main modes. For N = 1 it clearly has the greatest value, whereas for other electron numbers it does not vary much.

Figure9 shows non-interacting FIR spectra of square-symmetric four-minima QDM. In N = 3 spectrum of figure9(c) there is one additional mode below ω_- and one anti-crossing in ω_-.

The additional mode below ω_- is much closer to ω_- in energy as it is in the double dot spectrum. Also ω_- has lower transition probability than the additional mode in four-minima QDM. N = 4, 5 and 6 spectra of four-minima QDM show multiple modes of ω_+ and some structure in ω_-, but these are very close in energy and may not be resolved in experiments.

However, there is a discernible split-off mode below ω_+ mode with a lower transition probability. It is difficult to say how anti-crossing gaps of ω_+ at $B \approx$ 1.7T are changing with increasing N because of multiple modes. One interesting thing is that zero-field excitation of the degenerate main branches is higher, $E \approx$2.6meV, with N = 3-6 than it is with N = 1 and 2, where $E \approx$2.1meV.

Non-interacting FIR spectra of rectangular-symmetric four-minima QDM in figure10 does not deviate significantly from the FIR spectra of two other QDM confinements. Additional mode below ω_- is very weak but visible and exist only at very low magnetic field strengths in rectangular-symmetric four-minima QDM for $N = 3$, 4 and 6 in figure10. Clear anti-crossing in ω_- occurs only with $N = 3$. The zero-field gap between ω_+ and ω_- becomes very small with more than one electron in the QDM. After the first anti-crossing, at $B \approx 2$T, the ω_+ is split to two modes with $N \geq 1$, where the lower mode has lower transition probability.

In the calculated FIR spectra we are able to compare three different QDM confinement potentials and study the effects of interactions on the two-electron FIR spectra. We have also calculated FIR spectra for a few non-interacting electrons. The three studied lateral QDM confinements are a two-minima QDM (double dot), and square-symmetric and rectangular-symmetric four-minima QDMs.

We observe anti-crossings in the upper ω_+ mode for both interacting and non-interacting spectra. The anti-crossing gaps are not significantly altered by the interactions, but the anti-crossing positions are shifted to lower magnetic field values with the interacting electrons. When we increase the particle number in the non-interacting FIR spectra, the anti-crossings in ω_+ modes become less clear as the main mode is split to multiple modes at the anti-crossing point. With three non-interacting electrons ($N = 3$) we see a clear anti-crossing also in the lower ω_- mode in all three QDM confinement potentials.

The ω_+ mode is split to two for N e" 2 non-interacting electrons and also in the interacting spectra when the spin triplet is the ground state.

With two interacting electrons this additional mode vanishes from the spectrum at greater magnetic field values, approximately at B e" 4 for all QDM confinements. The upper split-off mode has a lower transition probability in the interacting case in double dot and rectangular four-minima QDM confinements. In the non-interacting case, even if we add electrons up to six electrons in a QDM, the ω_+ consists of only two energy-split excitations, where one of the modes, typically the lower one, can have smaller transition probability in all three QDM confinements.

The third point that we have analysed in the FIR spectra

is the splitting between the main modes, $\omega_{\pm}$, at $B = 0$. The splitting is seen for two rectangular confinements, double dot and rectangular-symmetric four-minima QDM, where the excitations along short and long axes are not equal. In the two rectangular confinements, the greatest gap is seen for single electron.

For all $N \geq 2$ the gap is clearly smaller compared to $N = 1$ but does not change significantly between N = 2-6 non-interacting electrons in the QDM. The zero-field splitting usually diminishes as electron-electron interactions effectively steepen the potential which leads to increase in excitation energies. Interactions have stronger effect on the lower mode, which corresponds to excitation along the long axis of the confinement, therefore decreasing the gap.

Generally, QDM confinements have lower excitation energies compared to a parabolic QD with the same confinement energy $\hbar\,\omega_0$. When interactions are turned on, the Coulomb repulsion effectively steepen the potential leading to a small increase in the excitation energies.

The same features are in general present in interacting and non-interacting spectra, but small shifts may occur with interacting electrons compared to the non-interacting FIR spectra.

Therefore, based on calculations with two interacting electrons, we conclude that deviations from the Kohn modes are due to the non-parabolic confinement potential. Interactions only shift some features in the observed spectra but otherwise excitations are of a collective nature. However, this study is only for two interacting electrons and maybe more features can be seen with more interacting electrons in a non-parabolic QD confinement. Also it is possible, but unlikely, that different non-parabolic confinement potentials could show more interaction effects in FIR spectra.

We have studied how the FIR spectra change when interactions are turned on and off. These predictions could be checked in experiments when the number of electrons is increased in the dots, even if the limit of single or two electrons could not be achieved. These predictions are that (i) anti-crossings are shifted to lower magnetic field values, (ii) excitation energies increase and (iii) zero field gaps decrease as the Coulomb interaction increases in the QD (more electrons inside dots).

It is also possible that the confinement potential may change as the number of electrons changes in the dot. In a rectangular symmetry also polarized light can be used to analyse excitations. These features are seen, in general terms, in elliptic QDs studied by Hochgrδfe *et al* . However, they observe a little more complicated FIR spectra with multiple negative and positive dispersion Kohn modes.

Chapter 13

Lattice Specific Heat of Carbon Nanotubes

Carbon nanotubes are nanostructures with remarkable electronic and mechanical properties. As soon as they were discovered, their potential applications stimulated several theoretical and experimental studies. Electronic and lattice properties, such as specific heat and thermal transport coefficients, in carbon nanotubes [1-6] and nanowires have been deeply investigated. Measurements with microfabricated devices show very interesting thermal transport properties in nanosystems . Thermal conductivity increases as the tube's diameter decreases. At low temperatures, thermal conductance of carbon nanotube bundles follows the power law T^a, where T is the temperature, with an exponent a less than 2 (around 1.5).

This fact suggested that the thermal transport in the bundles is like that of a quasi one-dimensional system. The same for specific heat: the measured specific heat differs from that of graphene and graphite, especially in the low temperature region, where quantization of the phonon band structure is observed. The more evident property of thermal conductivity in carbon nanotubes is its increase as the carbon nanotube diameter descreases. A totally different behaviour is observed in nanowires, where the thermal conductivity is reduced if the wire diameter is small. In nanowires, exponent a in the power law T^a ranges from 1 to 3, increasing with the wire section.

Here, in the framework of a recently proposed lattice

model , we will discuss the thermal properties of the phonon system displayed by a single wall carbon nanotube in the armchair configuration. In Ref. , the authors deeply investigate the role of symmetry rules, pointing out that the violation of these rules destroys the flexural modes. They did not discuss the lattice thermal properties of the model. Here, the specific heat is estimated to see if the Mahan and Jeon model gives a behaviour at low temperature in agreement with experimental data. A discussion on thermal conductivity follows.

PHONONS

A single wall carbon nanotube is a sheet of graphene rolled up along the line connecting two lattice points, into a seamless cylinder. According to the chosen line, a carbon tube can be achiral or chiral. The seamless condition brings to continuity conditions on tube surface for the phonon vibrations and for the electron distribution. Several references discuss the behaviour of electrons and electronic transport in nanotubes and phonon dispersions. Here, to study the lattice thermal properties, we follow the lattice model proposed by G.D. Mahan and G.S. Jeon in Ref.

These authors considered a "spring and mass" model with first and second neighbour bonds, and introduced a radial bond bending term in the potential, to include the curvature effect of the surface. The potential they propose satisfies the symmetry rules imposed by the cylindrical geometry. An armchair ?*n, n*? tube with diameter R, has *n* atoms A and *n* atoms B along the circumference: each carbon atom A (or B) is connected with three nearest neighbour atoms B(A) at a distance *a.*

Then, two integer numbers are required to describe the lattice sites (?*l, m)*?. *l* is denoting the position of lattice sites in the direction of the nanotube axis and *m,* ranging from 0 to *n* ?1, gives the positions of sites A and B on the circumference. Phonons in nanotubes have then two quantum numbers, one from the quantization of wave vectors in the direction of the tube axis, the other due to a quantization around the circumference. Let us call *L* the length of the tube, *N* the total number of lattice points, M the atomic mass and ω the angular frequency for a given wave vector q_z along the axis direction, for given values of polarization p and the quantum parameter n.

Phonon dispersions are evaluated. Oscillations of the lattice sites are traveling along the z -axis and, as found with the continuum models of hollow tubes, four acoustic modes are displayed. Two of them are the twist mode in tangential direction and the longitudinal stretching mode in the z -direction.

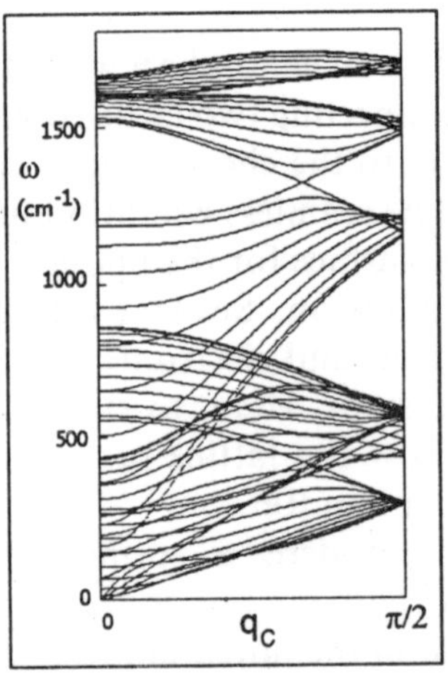

Fig. Phonon Frequency ω as a Function of Wave number q_z for an Armchair Tube (10, 10).

The other two acoustic modes are the flexural oscillations. Flexural modes are also observed in wires . The lower breathing mode in radial direction has a finite frequency for q_z =0, due to the lattice curvature. Three interactions, representing first neighbour, second neighbour and radial bondbending forces, are used in this lattice model: their contributions are weighted with relative coefficient r_j (r_1 = 1.0, r_2 = 0.06), r_3 = 0.024 respectively.

Figure show the angular frequency for an armchair (10, 10) carbon nanotube, as a function of the reduced wavenumber cq_z,where parameter c is equal to $\sqrt{3}a/2$. Here we use a scale factor Φ_0 for the potential $\left(\Phi_o = 4\times10^{20}\, g\, cm^{-2} s^{-2}\right)$, the value of which is chosen to give an acoustic longitudinal velocity of 16.7 km/s and an acoustic twist velocity of 9.1 km/s. The lattice constant is a = 1.42 × 10^{-8} cm.

LATTICE SPECIFIC HEAT

The lattice heat capacity is given by the following expression: $c = \frac{\hbar^2}{k_B T^2} \sum_{p.n} \frac{L}{2\pi} \int dq\omega^2 b_o(1+bo)$ and then the specific heat per atom is obtained, dividing by the number of the lattice sites. If we imagine a very long nanotube, the integral in equation is a good substitute for the sum on all the values of the lattice wavenumber along the *z* - direction. Sum on polarization and eigenvalue number *n* remains. *bo* is the unperturbed phonon distribution.

Figure shows the behaviour of the specific heat as a function of the temperature for three armchair tubes, (5,5), (10,10) and (20,20), in comparison with the experimental data [1,6]. As observed by V.N. Popov , two distinct behaviors at low and high temperatures are clearly displayed by tube (5,5) and (10,10): they are due to phonon frequency quantization.

HEAT TRANSPORT

Phonon heat transport can be easily approached in the framework of the time relaxation approximation. The thermal conductivity due to phonons is linked to the perturbed phonon distribution *b*: $b - b_o = -\psi \frac{\partial b_o}{\partial(\hbar\omega)}$ where *bo* is the equilibrium phonon distribution and Ψ a deviation function. The linearized Boltzmann equation for a solid subjected to a thermal gradient, written in the relaxation time approximation. where v is the phonon group velocity and t a phonon relaxation time. equation is a very rough estimate of Ψ.

The heat current density can be defined, for a nanotube with volume Ω = *SL* where *S* is the section and *L* the length, as: $U = -\frac{1}{2\pi S}\sum_{p,n}\int dq_z\, \hbar\omega v\psi \frac{\partial b_o}{\partial(\hbar\omega)}$ where the sum on all the lattice wavenumbers along the z - direction is evaluated with an integral.

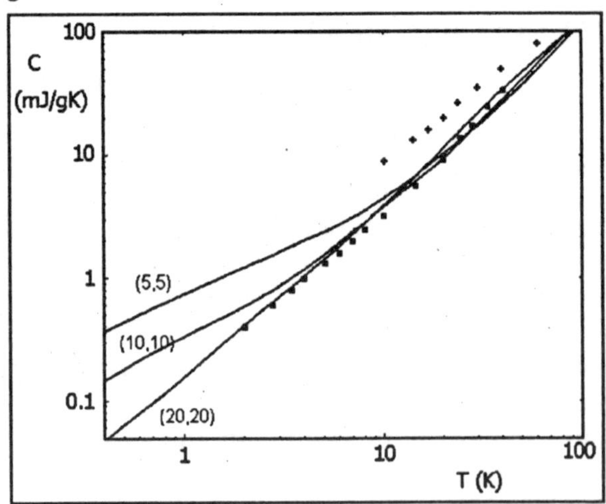

Fig. Specific Heat as a Function of Temperature for Single Wall Armchair Nanotubes (5, 5), (10, 10), (20, 20), Compared with Experimental Data.

A thermal current exists along the tube z-axis, if the nanotube is subjected to a gradient $\Delta T = \partial T/\partial z$ v is the phonon velocity in the z -direction. The thermal conductivity is defined as that parameter κ joining the heat current with the thermal gradient: $U = \kappa \partial T/\partial z$. If the phonons are only subjected to boundary scattering, that is to the scattering occurring at the ends of the tube, the relaxation time is given by $\tau = L/v$. Thermal conductivity turns out to be:

$$\kappa = \frac{\hbar^2 a^3 L}{2\pi k_B T^2 S}\left(\frac{\Phi_o}{M}\right)^{3/2} \times \sum_{p,n}\int d\varsigma \bar{\omega}^2 \bar{v} b_o (1 + b_o)$$

where

$$\omega = a\sqrt{\frac{\Phi_o}{M}}\bar{\omega}; \varsigma = cq_z; \bar{v} = |\partial\bar{\omega}/\partial\varsigma|$$

Equation is the kinetic model of thermal conductivity, if v is assumed equal to the mean value of the phonon velocity. To compare with experimental data reported and obtained measuring the thermal conductivity of single carbon nanotubes, integer n it put equal to 150: then the armchair tube diameter is 14 *nm,* as in experiments. To have an agreement with experimental data, the nanotube length L must be equal to 6.2 *μm.* Thermal conductivity from equations and experimental data are shown in figure. The three curves in this figure are given for different values of tube diameter: thermal conductivity is strongly increased by reducing the diameter.

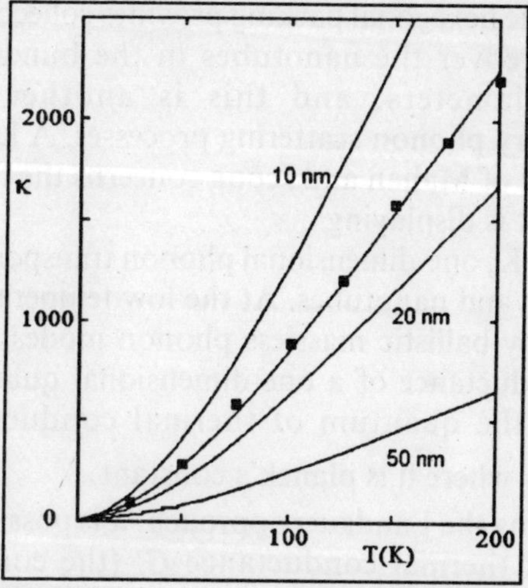

Thermal conductivity (in *W/m. K*) as a function of temperature for a single wall nanotube with diameter 14 *nm.* Curves refer to different values of the diameter. Once more: this is only a very rough estimate of the thermal transport. A rigorous calculation of thermal conductivity must consider normal and umklapp three-phonon scatterings and more details on phonon boundary scattering: both issues are very hard to discuss with a lattice model. Of course, MD simulations can give better performances in a more rigorous framework.

Curves in figure are following law T^a in the temperature range between 20 and 40 K, with $a = 1.75$. Other experimental data have been collected from bundles containing several nanotubes, all with parallel axes, in the hexagonal packing. Two bundles formed with tubes of different diameter (10 *nm* and 148 *nm*) have the experimental outputs of Ref. In this reference, a behaviour T^a with a = 1.5, is deduced from the experimental points. Thermal conductivity is strongly reduced in the nanotube bundles, if compares with the values obtained in measurements on single nanotubes.

With equations to obtain the experimental values of Ref.4, we must use a very short mean free path L of about 25 *nm*. Of course, if the nanutubes are packed in boundles, new phonon dispersions and phonon scattering mechanisms arise. Ref.6 shows how the hexagonal packing provides collective vibration modes. Moreover the nanotubes in the bundle can have different diameters, and this is another source of supplementary phonon scattering processes. A last comment on the model of Mahan and Jeong concerns the four acoustic modes that it is displaying.

Below 1 K, one-dimensional phonon transport is possible, in nanowires and nanotubes. At the low temperature regime, dominated by ballistic massless phonon modes, the phonon thermal conductance of a one-dimensional quantum wire is quantized, the quantum of thermal conductance being $\pi^2 k_B^2 T/(3h)$ where h is planck's constant.

Following the Landauer approach, it is possible to obtain the ballistic thermal conductance G, (the conductance G obtained for a diffusive regime fails at very low temperature). The Landauer approach gives the thermal conductance as : $G = g\pi^2 k_B^2 T/(3h)$ where g is the number of acoustic phonons. Acoustic modes are the channels through which the heat is transferred through a bridge (nanotube or nanowire) from a phonon reservoir to another. Any model of the brigde, nanotube or nanowire, must then involve no less than four acoustic modes.

Chapter 14

Nanoelectronics

Nanoelectronics refer to the use of nanotechnology on electronic components, especially transistors. Although the term *nanotechnology* is generally defined as *utilizing technology less than 100 nm in size*, nanoelectronics often refer to transistor devices that are so small that inter-atomic interactions and quantum mechanical properties need to be studied extensively.

As a result, present transistors (such as recent Intel Core i7 Processors from Intel) do not fall under this category, even though these devices are manufactured under 65nm or 45nm technology.

Nanoelectronics are sometimes considered as disruptive technology because present candidates are significantly different from traditional transistors. Some of these candidates include: hybrid molecular/semiconductor electronics, one dimensional nanotubes/nanowires, or advanced molecular electronics.

The sub-voltage and deep-sub-voltage nanoelectronics are specific and important fields of R & D, and the appearance of new ICs operating almost near theoretical limit (fundamental, technological, design methodological, architectural, algorithmic) on energy consumption per 1 bit processing is inevitable.

The important case of fundamental ultimate limit for logic operation is reversible computing.

Although all of these hold immense promises for the future, they are still under development and will most likely not be used for manufacturing any time soon.

APPROACHES TO NANOELECTRONICS

NANOFABRICATION

For example, single electron transistors, which involve transistor operation based on a single electron. Nanoelectromechanical systems also falls under this category.

Nanofabrication can be used to construct ultradense parallel arrays of nanowires, as an alternative to synthesizing nanowires individually.

Nanomaterials Electronics

Besides being small and allowing more transistors to be packed into a single chip, the uniform and symmetrical structure of nanotubes allows a higher electron mobility (faster electron movement in the material), a higher dielectric constant (faster frequency), and a symmetrical electron/hole characteristic. Also, nanoparticles can be used as quantum dots.

Molecular Electronics

Single molecule devices are another possibility. These schemes would make heavy use of molecular self-assembly, designing the device components to construct a larger structure or even a complete system on their own. This can be very useful for reconfigurable computing, and may even completely replace present FPGA technology.

Molecular electronics is a new technology which is still in its infancy, but also brings hope for truly atomic scale electronic systems in the future. One of the more promising applications of molecular electronics was proposed by the IBM researcher Ari Aviram and the theoretical chemist Mark Ratner in their 1974 and 1988 papers *Molecules for Memory, Logic and Amplification*, (see Unimolecular rectifier). This is one of many possible ways in which a molecular level diode / transistor might be synthesized by organic chemistry. A model system was proposed with a spiro carbon structure giving a molecular diode about half a nanometre across which could be connected by polythiophene molecular wires. Theoretical calculations showed the design to be sound in principle and there is still hope that such a system can be made to work.

Other Approaches

Nanoionics studies the transport of ions rather than electrons in nanoscale systems.

Nanophotonics studies the behaviour of light on the nanoscale, and has the goal of developing devices that take advantage of this behaviour.

NANOELECTRONIC DEVICES

Radios

Nanoradios have been developed structured around carbon nanotubes.

Computers

Nanoelectronics holds the promise of making computer processors more powerful than are possible with conventional semiconductor fabrication techniques. A number of approaches are currently being researched, including new forms of nanolithography, as well as the use of nanomaterials such as nanowires or small molecules in place of traditional CMOS components. Field effect transistors have been made using both semiconducting carbon nanotubes and with heterostructured semiconductor nanowires.

Energy Production

Research is ongoing to use nanowires and other nanostructured materials with the hope to create cheaper and more efficient solar cells than are possible with conventional planar silicon solar cells. It is believed that the invention of more efficient solar energy would have a great effect on satisfying global energy needs.

There is also research into energy production for devices that would operate *in vivo*, called bio-nano generators. A bio-nano generator is a nanoscale electrochemical device, like a fuel cell or galvanic cell, but drawing power from blood glucose in a living body, much the same as how the body generates energy from food. To achieve the effect, an enzyme is used that is capable of stripping glucose of its electrons, freeing them for use in electrical devices. The average person's body could, theoretically, generate 100 watts of electricity (about 2000 food calories per day) using a bio-nano generator. However, this estimate is only true if all food was converted to electricity, and the human body needs some energy consistently, so possible power generated is likely much lower. The electricity generated by such a device could power devices embedded in the body (such as pacemakers), or sugar-fed nanorobots. Much of the research done on bio-nano generators is still experimental, with Panasonic's Nanotechnology Research Laboratory among those at the forefront.

Medical Diagnostics

There is great interest in constructing nanoelectronic devices that could detect the concentrations of biomolecules in real time for use as medical diagnostics, thus falling into the category of nanomedicine. A parallel line of research seeks to create nanoelectronic devices which could interact with single cells for use in basic biological research. These devices are called nanosensors. Such miniaturization on nanoelectronics towards in vivo proteomic sensing should enable new approaches for health monitoring, surveillance, and defence technology.

MOLECULAR ELECTRONICS

Molecular electronics (sometimes called *moletronics*) is an interdisciplinary theme that spans physics, chemistry, and materials science. The unifying feature of this area is the use of molecular building blocks for the fabrication of electronic components, both passive (e.g. resistive wires) and active (e.g. transistors). The concept of molecular electronics has aroused much excitement both in science fiction and among scientists due to the prospect of size reduction in electronics offered by molecular-level control of properties. Molecular electronics provides means to extend Moore's Law beyond the foreseen limits of small-scale conventional silicon integrated circuits. Due to the broad use of the term, molecular electronics can be split into two related but separate subdisciplines: *molecular materials for electronics* utilizes the properties of the molecules to affect the bulk properties of a material, while *molecular scale electronics* focuses on single-molecule applications.

CONCEPT GENESIS AND THEORY

Study of charge transfer in molecules was advanced in the 1940s by Robert Mulliken and Albert Szent-Gyorgi in discussion of so-called "donor-acceptor" systems and developed the study of charge transfer and energy transfer in molecules. Likewise, a 1974 paper from Mark Ratner and Ari Aviram illustrated a theoretical molecular rectifier. Later, in 1988, Aviram described in detail a theoretical single-molecule field-effect transistor. Further concepts were proposed by Forrest Carter of the Naval Research Laboratory, including single-molecule logic gates.

These were all theoretical constructs and not concrete devices. The *direct* measurement of the electronic characteristics of individual molecules awaited the development of methods for making molecular-scale electrical contacts. This was no easy task. Thus, the first experiment measuring the conductance of a single molecule was only reported in 1997 by Mark Reed and co-workers. Since then, this branch of the field has progressed rapidly. Likewise, as it has become possible to measure such properties directly, the theoretical predictions of the early workers have been mostly confirmed.

Voltage-controlled switch, a molecular electronic device

from 1974. However, while mostly operating in the quantum realm of less than 100 nanometers, "molecular" electronic processes often collectively manifest on a macro scale. Examples include quantum tunneling, negative resistance, phonon-assisted hopping, polarons, and the like.

Thus, macro-scale active organic electronic devices were described decades before molecular-scale ones. E.g., in 1974, John McGinness and his coworkers described the putative "first experimental demonstration of an operating molecular electronic device". This was a voltage-controlled switch. As its active element, this device used DOPA melanin, an oxidized mixed polymer of polyacetylene, polypyrrole, and polyaniline. The "ON" state of this switch exhibited almost metallic conductivity.

Since the 1970's, scientists have developed an entire panoply of new materials and devices. These findings have opened the door to plastic electronics and optoelectronics, which are beginning to find commercial application.

CHARGE TRANSFER COMPLEXES

The first highly-conductive organic compounds were the Charge transfer complexes. In 1954, researchers at Bell Labs and elsewhere reported Charge transfer complexes with resistivities as low as 8 ohms-cm. In the early 1970's, salts of tetrathiafulvalene were shown to exhibit almost metallic conductivity, while superconductivity was demonstrated in 1980. Broad research on charge transfer salts continues today.

Conducting Polymers

The linear-backbone "polymer blacks" (polyacetylene, polypyrrole, and polyaniline) and their copolymers are the main class of conductive polymers. Historically, these are known as Melanins. In 1963 Australians DE Weiss and coworkers reported iodine-doped oxidized polypyrrole blacks with resistivities as low as 1 ohm/cm. Subsequent papers reported resistances as low as 0.03 Ohm/cm. With the notable exception of Charge transfer complexes (some of which are even superconductors), organic molecules had previously been considered insulators or at best weakly conducting semiconductors.

Over a decade later in 1977, Shirakawa, Heeger, and MacDiarmid reported equivalent high conductivity in rather similarly oxidized and iodine-doped polyacetylene. They later received the 2000 Nobel prize in chemistry for "The discovery and development of conductive polymers". The Nobel citation made no reference to Weiss *et al's* similar earlier work. Also see Nobel Prize controversies.

C_{60} AND CARBON NANOTUBES

From graphite to C_{60}

In polymers, classical organic molecules are composed of both carbon and hydrogen (and sometimes additional compounds such as nitrogen, chlorine or sulphur). They are obtained from petrol and can often be synthethized in large amounts. Most of these molecules are insulating when their length exceeds a few nanometers. However, naturally occurring carbon is conducting. In particular, graphite (recovered from coal or encountered naturally) is conducting. From a theoretical point of view, graphite is a semi-metal, a category in between metals and semi-conductors. It has a layered structure, each sheet being one atom thick. Between each sheet, the interactions are weak enough to allow an easy manual cleavage.

Tailoring the graphite sheet to obtain well defined nanometer-sized objects remains a challenge. However, by the close of the twentieth century, chemists were exploring methods to fabricate extremely small graphitic objects that could be considered single molecules. After studying the interstellar conditions under which carbon is known to form clusters, Richard Smalley's group (Rice University, Texas) set up an experiment in which graphite was vaporized using laser irradiation. Mass spectrometry revealed that clusters containing specific "magic numbers" of atoms were stable, in particular those clusters of 60 atoms. Harry Kroto, an English chemist who assisted in the experiment, suggested a possible geometry for these clusters - atoms covalently bound with the exact symmetry of a soccer ball. Coined buckminsterfullerenes, buckyballs or C_{60}, the clusters retained some properties of graphite, such as conductivity. These objects were rapidly envisioned as possible building blocks for molecular electronics.

CARBON NANOTUBES

THEORY OF MOLECULAR ELECTRONICS

The theory of single molecule devices is particularly interesting since the system under consideration is an open quantum system in nonequilibrium (driven by voltage). In the low bias voltage regime, the nonequilibrium nature of the molecular junction can be ignored, and the current-voltage characteristics of the device can be calculated using the equilibrium electronic structure of the system. However, in stronger bias regimes a more sophisticated treatment is required, as there is no longer a variational principle. In the elastic tunneling case (where the passing electron does not exchange energy with the system), the formalism of Rolf Landauer can be used to calculate the transmission through the system as a function of bias voltage, and hence the current. In inelastic tunneling, an elegant formalism based on the non-equilibrium Green's functions of Leo Kadanoff and Gordon Baym, and independently by Leonid Keldysh was put forth by Ned Wingreen and Yigal Meir. This Meir-Wingreen formulation has been used to great success in the molecular electronics community to examine the more difficult and interesting cases where the transient electron exchanges energy with the molecular system (for example through electron-phonon coupling or electronic excitations).

Nano-Electronics

Todays micro/nano-electronics is characterized by *migration of research from pure down-scaling to new functionality and combined technology-system innovation.* This is mainly required in order to manage power dissipation, variability and complexity issues that are associated with tera-systems that exploit nano-technology. While this new functionality and technology-system innovation will exploit the properties of future nano-scaled technologies, the aggressive scaling will also play a fundamental, but not an exclusive role. New research drivers, such as ultra-low power technologies, bio-inspired circuit and system design and ambient intelligence applications are expected to play an increasingly important role in the next years.

The generally accepted view of micro and nano-electronics identifies three main research domains: (i) *More Moore*, (ii) *Beyond CMOS* and (iii) *More than Moore*.

The More Moore domain is essentially dealing with technologies related to the nanometer CMOS; prevailing business directions together with continuing scalability have determined the evolution of silicon CMOS as the technology with the highest added value. In Nano-Tera.CH, CMOS technology will be considered as a key platform to supply the massive computing power and communication capability needed for the realization of Ambient Intelligence applications at an affordable cost and a power efficiency exceeding todays leading-edge examples.

The Beyond CMOS domain deals with disruptive technology and device principles (from charge to non-charge based devices, from semiconductor to molecular technology), compared to CMOS, to exploit atomic-scale technology and novel functionality. Novel switches, architectures for universal memory and new interconnect approaches are some of the identified challenges. In addition to technologies such as nanowires and nanotubes, particular attention will be devoted to the development of disruptive technologies such as molecular electronics, over a longer time horizon. In this perspective, partial (hybridization) or total replacement solutions for silicon CMOS are foreseen.

The More than Moore field is of strategic relevance for Nano-Tera.CH; it deals with true engineering of complex systems that combine, by heterogeneous integration techniques, various technologies. The More than Moore technology platform also gives particular attention to MEMS/ NEMS and sensor technologies that are combined to silicon or non-silicon computing, information storage and encryption and wireless / RF communication technologies.

Major micro/nano-electronic research priorities for Nano-Tera.CH are organized and discussed in three sections:

- Technology (top down and bottom up)
- Devices and Circuits
- Systems and Integration.

Technology

Within the last decade, the nanotechnology and nanodevice evolutions have been dominated by the aggressive scaling in the sub-100nm zone and the related cost-effectiveness of silicon technology. Today top-down nanolithographic based approaches are challenged by bottom-up nanotechnology; Nano-Tera.CH considers that a smart combination and developments of the two approaches will be needed in the future and the choices will be dictated by complex system performance and functionality. *Atomic-scale technologies* with *nano-to-micro interface technologies* and the possible *extension of the Moores law in the 3rd dimension* for building tera-systems are ultimate goals of the fabrication research and development. Below, some of the most important topics in nanotechnology are discussed.

Top-Down (mostly silicon based): Presently, silicon top-down technology (Nanometer CMOS) is the single recognized solution for reliable circuits and systems. In order to meet the challenges of near-10nm gate length, the Nanometer CMOS will have to integrate the so-called technology boosters, such as high-k metal gate stacks and transport enhancement techniques (strained devices and bandgap engineering). Top-down etched silicon nanowires or the integration of III-V nanowires on silicon CMOS will then become alternative platforms on which 3D nano-processing (such as in the case of Gate-All-Around nanowire transistors), process control, abrupt junction (smaller than 2nm) and metal contacting will have to deal with tolerable variability and high density. Other top down techniques, such as nano-imprint and nano-stenciling will have to be considered as alternative to e-beam or other advanced lithography.

Bottom-Up: Bottom up nanotechnology is a unique opportunity for nanoelectronics to develop and exploit self-assembly approaches, which could offer unrivaled density of functionality beyond lithographic limits. In the this context, the exploration of both organic and inorganic materials, the nanowire and nanotube growth and self-assembling techniques of these one dimensional structures, the possibility offered by semiconducting or metallized DNA technology and the controlled fabrication and exploitation of 0-dimensional structures (nanodot technology) are clear identified priority domains. In general, molecular electronic technologies that answer the computation, memory and interconnectivity challenges to build and serve for true tera-scale systems, will be considered as future alternative of silicon CMOS, under the nanoelectronics umbrella.

On top of those, other realistic emerging fabrication techniques and related materials for large area electronics, where the global dimensionality of the system is completely different, such as direct printing techniques and ink-jetting (using semiconducting and conducting inks based on nanoparticles), with applications in flexible electronics, or enabling advanced concepts as invisible or paint-able electronics, will be considered.

Devices and Circuits

Novel devices and circuit architectures, truly exploiting the new emerging technology are key aspects of Nano-Tera.CH research because of their impact in all application domains and of the possibility to create unique know-how around the new concepts. While in the digital domain the world-wide research is dominated by the quests for the new quasi-ideal switch and the so-called universal memory (a memory concept that is non-volatile and able to integrate all the microprocessor memory hierarchy, including the slow external drive, on a single dense and high-speed memory platform), Nano-Tera.CH will also consider, on top of those, the device quest for new functionality and ultra-low power.

Ultra low-power devices and circuits exploiting both solid-state but also hybrid principles (where electronics is combined with micromechanics, optics, spin devices, etc) will enable standby power levels that are compatible to the practical requirements of the wireless sensor nodes (sub-100microW/node) and the true possibility of considering energy scavengers at the scales of power. New switch devices such as tunnel FET, NEM-FET, IMOS or, at much longer term, some more exotic device architectures exploiting discrete charge, such as Single Electronics or QCA-like, could play a significant role in this context.

Finally, it is worth mentioning that, in general, individual molecular devices appear promising for future high-dense ultra-low power electronics; particular attention will be paid to the demonstration of novel circuit and system architectures exploiting the unique molecular device features (for instance, the possible use of molecular switches in cross-bar memory architecture, enabling an increase of memory density by 102 compared to any other conventional approach).

High-speed and telecom devices and circuits operating at frequencies in excess of hundreds of GHz, and approaching the THz regime, will be essential for enabling communication interfaces and More-than-Moore system solutions. Co-design of analog ICs with RF MEMS devices (for tunable/programmable RF front-ends and also, including antenna technology) will be there essentially for achieving both high frequency (HF) and low power operation. These new devices are expected to open an era of electro-mechanical signal processing and, especially, HF system reconfigurability.

A particularly important category is the *bio-inspired devices and circuits*; in this direction, priorities will be given to the increase of the local inter-connectivity of individual devices, such as in the neuron-inspired devices case and the processing of the information using principles other than the binary logic (analog or weighted multiple input approaches; neural-inspired circuits, in general). In the case of these architectures, the targeted performance factors will be not primarily related to high speed but rather parallel processing of information at extremely low power and 3D device and circuits architectures.

Universal memory device architectures and related addressing circuits is so far identified as a strategic research direction in micro/nanoelectronics; this is motivated by the fact that all the tera-system applications are requiring today huge amount of memory, as dense as possible and very high-speed. Moreover, today there is no clear winner in the quest for the universal memory; PCM/OUM (phase change/ovonic), FLASH, MRAM (magnetic RAM), FeRAM (Ferroelectric RAM), polymer memories, Millipede, etc., are all competing in an increasing market of non-volatile memories.

Disposable electronics and applications in monitoring (security) and the medical field are expected to be supported by the category of *flexible and large-area electronics.* Ultra-thin film organic and non-organic materials (integrable on flexible substrate) with high carrier mobility, low voltage operation, the control of device lifetime and of the drift of characteristics of these devices, their processing with ultra low temperature budget (less than 200C) and the co-integration of digital, analog, sensing, memory and communications functions on flexible substrates, all are challenges to be addressed in the future.

Systems and Integration

Innovative breakthrough ideas that enable true tera-scale system integration will play a central role in Nano-Tera.CH, opening up the possibility of achieving system complexities that are two-to-three orders of magnitude higher than todays state-of-the-art. This capability is absolutely key for the successful demonstration of the fabrication technologies and novel device/circuit concepts that are explored under the umbrella of the research initiative. Equally important, the systems integration aspect will be used as a touchstone for evaluating and for qualifying various nano-scale device technologies, in the sense of determining the system-ability of such candidate technologies.

3-dimensional (3D) and vertical integration of nanoelectronic devices, memory elements, and interconnect layers is considered as a very promising technique to increase the overall integration density as well as the performance of complex systems, mainly due to the significant shortening of connection paths between layers (strata). However, a number of very challenging issues will need to be resolved in order to allow widespread adoption of 3D integration for large-volume production: realization of high-density thru-wafer vias, multi-level wafer bonding techniques, addressing the thermal management issues for intermediate layers, development of effective 3D layer assignment, placement and 3D routing algorithms to mention a few. True 3D integration is expected to open the possibilities for very-small-form-factor multi-processing and memory units which can be used in wearable / ubiquitous systems as well as implanted devices. This 3D integration aspect will also become one of the central technology drivers of the Nano-Tera.CH initiative that will be utilized to merge many technology layers.

Heterogeneous integration platforms: In parallel to the development of nanometer CMOS as well as beyond-CMOS device technologies for switching, memory and analog functions, there is an increasing need to integrate various (heterogeneous) technologies without sacrificing the overall system performance and without further increasing the packaging costs. This may be accomplished by the judicious use of wafer bonding techniques, multi-chip packages, optical and electro-magnetic coupling between various components for high data transfer rates and improvement of SiP technologies. This becomes especially valuable when considering integrated MEMS/NEMS devices that must interface with the processing electronics in very close proximity, and in RF/wireless applications.

Reliability and fault tolerance of future complex systems built with nanoscale electronic devices cannot be guaranteed using conventional measures, due to fundamental physical limitations such as process variability, excessive leakage, process costs as well as very high power densities. This observation calls for radical action on several fronts in order to ensure the continuity of the nanoelectronic systems integration paradigm. In particular, under Nano-Tera.CH we will develop circuit-level measures to mitigate the limitations of process variations, leakage and reduced device reliability; and finally, explore system-level design approaches that are better adapted to the constraints imposed by the materials, technology, and device physics.

Thermal control / power-energy management: Power/energy dissipation of complex integrated systems continues to be the primary show-stopper at high speed, and also for wearable-implantable systems. The power/energy optimization of multi-core integrated systems will require continuous monitoring of the actual workload conditions and judicious choices for the operating voltage as well as the frequency of each computing element. Ground-breaking technological innovations (such as on-chip Peltier elements, or liquid cooling through micro-channels) as well as systematic implementation of aggressive power management techniques (such as on-line temperature monitoring/control and schedule optimization) will be needed to address these problems within the frame of the Nano-Tera.CH initiative. In particular, it is expected that future digital and mixed-signal circuits and systems will make extensive use of subthreshold (weak-inversion) mode operation, exploiting ultra-low-current techniques and massive parallelism (system-level multiplexing) to achieve high performance at a low power budget.

Wireless/RF systems: With ubiquitous and wearable systems as a main target area, it is envisioned that the low-power high data-rate wireless communication techniques and systems will play a decisive role for the realization of many demonstrator applications, within the frame of Nano-Tera.CH. Key issues that will be addressed include ultra-wide-band (UWB), extreme low-power RF front-end design for wearable applications, utilization of hybrid integration technologies such as RF-MEMS/NEMS components, wireless sensors, as well as RF-based energy transfer / energy scavenging.

Bio-medical interfaces / implantable systems: Development of efficient bio-medical interfaces involves the exploration of bio-compatible materials, possible utilization of nanoscale components such as CNTs, as well as ultra-low-power and ultra-low-noise electronics to sense and convert the biological signals without disturbing or interfering with the living tissue.

The developed technologies in this field are expected to be instrumental for a wide range of implanted and portable analytic devices, such as DNA and protein sensors, diagnostic Lab-on-a-Chip (LoC) devices and neuronal/electronic interfaces for therapeutic applications.

Optical interconnects / optical clock distribution: In high-density, high-performance digital systems, short distance (chip-to-chip and board-to-board) optical interconnects are driven by the need for larger bandwidth and better cross-talk immunity.

For on-chip optical interconnects, more fundamental challenges will have to be addressed such as the technological feasibility of silicon-based emitters and receivers, suitability of waveguide materials, and power dissipation limitations. Key challenges for optical clock distribution include the layout of a regular distribution network, placement of clock receivers/ repeaters, and the design of high-efficiency silicon photodetectors.

Computer-Aided design: Models, algorithms, tools and methodologies are needed to support experimental and production-level design with nano-technologies. Nano-Tera.CH will foster the developments of technologies for design (design technologies) that provide appropriate abstraction of the technology to support simulation (at various levels) of devices and circuits, as well as the synthesis of their physical design.

At the same time, design methods and tools to support logic design on nano-technologies, wit h specific attention to parallelism, redundancy and reliability issues.

It is worth mentioning that Nano-Tera.CH will pay particular attention to the benchmarking of new technology and devices according to given system level functionality and specifications; a so-called system-able micro/nano-electronics approach will be the key for success.

Some of the criteria for filtering out the true system-able nanotechnologies include: Ability to co-host/co-integrate three basic system functionalities: computation, storage and communication (interconnects); acceptable power density and/or energy consumption; device variability and cost-effective solutions; possibility to build analog or mixed signal systems from the new devices; reliability and yield adequate for envisioned applications; and the added value with respect to silicon CMOS.

Chapter 15

Nanoscale Imaging Technology For Thz-frequency Transmission Microscopy

This paper introduces a novel nanoscale-engineering methodology for the first-time development of a microscope-system capable of collecting terahertz (THz) frequency spectroscopic signatures from microscopic biological (bio) structures. Here, the main challenge is the realization of a very sensitive spectroscopic-sensing technique that can be applied successfully to measure (and map) the THz-absorption images produced when long-wavelength radiation is applied to bio-agents that possess distinct structural features at nanoscale dimensions and below.

This unique THz transmission microscopy is motivated primarily by previous studies on biological materials and agents that have produced spectral features within the THz frequency regime (i.e., ~ 300 GHz to 1000 GHz) that appear to be representative of the internal structure and characteristics of the biological samples – e.g., DNA, RNA and bacterial spores. However, these same studies (that utilized conventional apparatus and required difficult-to-control bulk samples for averaging over large apertures) have shown the serious need for new techniques that can extract spectral/spatial information from individual microstructures without the masking/variation effects associated with bulk macroscopic samples.1

Therefore, THz microscopy is the obvious solution for the effective probing of small bio-agents and the collection of *reproducible THz signatures* that have real value for detection applications. The nonexistence of a THz microscope is a simple manifestation of the difficulties (i.e., diffraction limit) involved in performing deep sub-wavelength (DS) imaging using conventional approaches. Specifically, if the traditional point of view is used, the resolution size is limited by the spot-size that can be produced by a Gaussian beam. However, this is not the only avenue to producing an image. In fact, DS imaging at optical and THz frequencies is an active area of research where a number of approaches (or tricks)2 are under investigation for significantly reducing the measurement aperture.

For example, both interferometric and near-field (nano-probe) techniques can realise nano-resolution, and it is possible to utilize a pump (short-wavelength) plus probe (longwavelength) method to affect the same result. However, all these methods effectively utilize a "microscopic" source (i.e., to realise a small aperture) and then collect the measured-energy with a "macroscopic" detector, and this severely limits sensitivity in spectroscopic (i.e., narrow bandwidth & small intensity) imaging applications. An obvious alternative approach is to employ a *"microscopic" detector* or imaging array (i.e., that would set small apertures) and a *"macroscopic" source* (i.e., that would provide for lots of power and increased sensitivity).

Of course this is very challenging from a technology perspective, but is exactly the one proposed here. Specifically, a novel quantum-cell (Q-cell) sensor concept (with nanoscale pixel resolution) is introduced that allows for sensing the intensity of THz photon-flux via their polarizing effect on singleelectrons which is combined with a proven RF single-electron-transistor (SET) based frequency-domain multiplexing approach to statistically readout the polarization-state of these electrons. As will be shown, when such Q-cell detection and RF sampling concepts are combined with advanced materials growth, fabrication & testing techniques, THzfrequency transmission microscopy becomes feasible such that spectral/spatial information can be acquired at the nanoscale. Hence, this proposed nanoscale imaging technology has potential for the study and characterization of microscopic biomaterials/ agents, and therefore has important relevance for defence and security applications against bio-threats.

In particular, the specific technology objective of this work is the identification of a very sensitive instrument that can be used to successfully measure spectroscopic absorption from bio-structures that have nanoscale spatial dimensions. Specifically, since the objects of interest have dimensions on the order of 100's of nanometers or less (which is much less than the spectral wavelength), this is equivalent to realizing THz-frequency deep sub-wavelength spectroscopic imaging (THz-DSSI).

Therefore, the goal is the design and demonstration of a *THz microscope* which is exceedingly difficult from a conventional point of view (i.e., since one is well below the diffraction limit). However, such a THz microscope offers significant potential payoffs for the scientific characterization of biomaterials/ agents and for the development of advanced sensor methodologies. Indeed, this technology development has been motivated because THz-frequency spectroscopy is potentially useful for detecting and identifying bio-agents, but important signature- phenomenology related problems must be solved before the technique can be made amenable to practical application. As will be discussed in the sections that follow, novel nanoelectronic-based technologies and methodologies offer promise for enabling THzsignature based detection and identification of bio-materials and bio-agents in the future.

SCIENTIFIC BACKGROUND

During the last ten years, spectroscopic measurements conducted on bio-materials/agents have produced spectral features within the THz frequency regime (i.e., ~ 300 GHz to 1000 GHz) that appear to be representative of the internal structure and characteristics the specific bio-systems under consideration – e.g., DNA, RNA and bacterial spores.1 However, the THz spectroscopic approach has been very problematic because the spectral features observed from bulk solid-state samples tends to be very weak (i.e., ~ 1- 5% local variation in spectral absorption) and of limited number within the band.

Also and very noteworthy to the THz sensing problem are recent results that suggest the underlying dynamics (i.e., vibrational modes) that produce the THz-frequency absorption characteristics might be seriously underestimated by measurements performed on large samples made up of microscopic bio-molecules or bioparticles (i.e., either ordered or random in nature). In particular, recent theoretical and experimental results suggest that the absorptions characteristics of isolated bio-systems may possess optically active vibrational (or phonon) modes that produce spectral absorption features that are much stronger and/or sharper (i.e., and therefore less overlapping between individual spectral signatures) that those obtained from measurements made on bulk samples.

Here it is important to note that bulk samples have been required in almost all the reported THz spectral studies to date because a sufficiently large and uniform aperture must be utilized in order to successfully measure the very weak phenomenon within the microscopic bio-system. Needless to say, one obviously needs extremely high sensitivity in performing such measurements and when large collections of either randomly placed or highly ordered bio-molecules are considered it becomes extremely difficult to execute repeatable spectroscopic experiments because it is difficult to prepare identical systems and to subsequently control them during testing.

Clearly, a fundamental approach for avoiding the previously discussed limitations is to prescribe a new technique whereby the THz signatures can be collected from individual bio-molecules, and this motivates the search for a viable THz microscopy technology as is presented here. From the usual perspective, it might at first seem somewhat foolish to image with long wavelength radiation due to the diffraction limit – i.e., if one wants more resolution for contrast imaging then a shorter wavelength is always better. However, the specific goal here is to collect spectral signatures within the THz regime and to do this for very small target objects.

This being accepted, it should not be surprising that there are existing approaches for addressing this problem. Indeed, one could utilize interferometric imaging to significantly reduce resolution as this has been very effective in nanoscale fabrication, and alternatively, there are many groups actively utilizing near-field nano-probe based techniques to reduce the available aperture. Apparently, both of these approaches can significantly benefit from pump-probe type operations, so it is certainly reasonable to attack the THz microscope problem using these techniques. Although, there are noteworthy challenges associated with the very small effective apertures. Furthermore, it can be expected that there will be definite sensitivity limits associated with these methods when spectroscopic operation is desired.

As noted earlier, both of the previous approaches effectively utilize a "microscopic" source (i.e., to realise a small aperture) and then collect the measurement energy with a "macroscopic" detector. A somewhat more obvious approach that could lead to successful microscopy would be to employ a "microscopic" detector or imaging array (i.e., to set a small aperture) and a "macroscopic" source (i.e., to provide more power and increased sensitivity). Of course, this is very challenging from a technology perspective and is usually dismissed in any discussion on the subject at hand. Indeed, if one seeks to achieve THz microscopy using a microscopic detection methodology then nanotechnology innovations of the type to be discussed here will be critical.

THZ MICROSCOPE CONCEPT AND SUPPORTING DEVICE TECHNOLOGY

Recent research by our collaborative group have focused on the preliminary design and analysis steps that will provide the basic guidelines for the future fabrication and demonstration of novel nanoelectronic devices and circuits that can be used to rapidly measure incident THz radiation with nanoscale spatial resolution (or pixel sizes) and with very high sensitivity (i.e., approaching the quantum limit). The specific focus to date has been on the *detector elements* and *readout circuitry* that can be used to perform THz- DSSI on microscopic bio-agents (e.g., bio-cells) with nanoscale structural features. This THz Transmission Microscope concept utilizes quantum-cell detectors (QCDs) that exhibit a polarization response to select THz radiation, along with RF single-electrontransistor circuitry that can rapidly measure these polarization changes (to infer incident energy flux).

A hypothetical implementation of the full-array microscope architecture is illustrated in figure, where QCDs (i.e., the green cubes) are designed to respond to narrow-band THz radiation through changes to the spatial-state polarizations of trapped electrons (see "right-" & left-pointing" arrows).

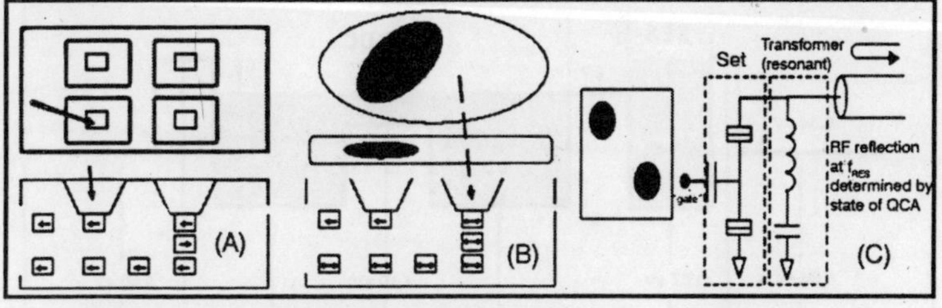

Fig. Detector elements and readout circuitry. (a) Hypothetical implementation of the full-array microscope architecture; (b) THz radiation induced electron polarization and the scheme of communication for future integration of the arrays; (c) RF measurement of the polarization of the trapped electron in the detection cell.

In this example, the scheme has been made compatible to Quantum Cellular Automata (QCA) communication for facilitating the future integration of the array. Architectures such as this could be used for transmission-mode microscopy of the type illustrated in figure. Here, nano-sized detectors (10-100 nm's) allow for measuring the presence of nano-features (e.g., virus) that might absorb higher levels of the THz radiation– i.e., as indicated by the dark red spot. Clearly, this is a very feasible approach to THz-DSSI if one can define

QCD's that *respond statistically* to the number of collected THz photons (incident intensity) and if the associated polarization processes are amenable to electrical measurement (i.e., *long time constants*) by known techniques for the RF measurement of the QCA state such as illustrated in figure.

A summary of the technology elements and proposed measurement scheme will now be given to illustrate the basic approach.

QUANTUM-CELL DETECTOR (QCD) CONCEPT

The basic approach presented here for achieving THz microscopy leverages singleelectron transition-processes and their dependence to long and short wavelength radiation.

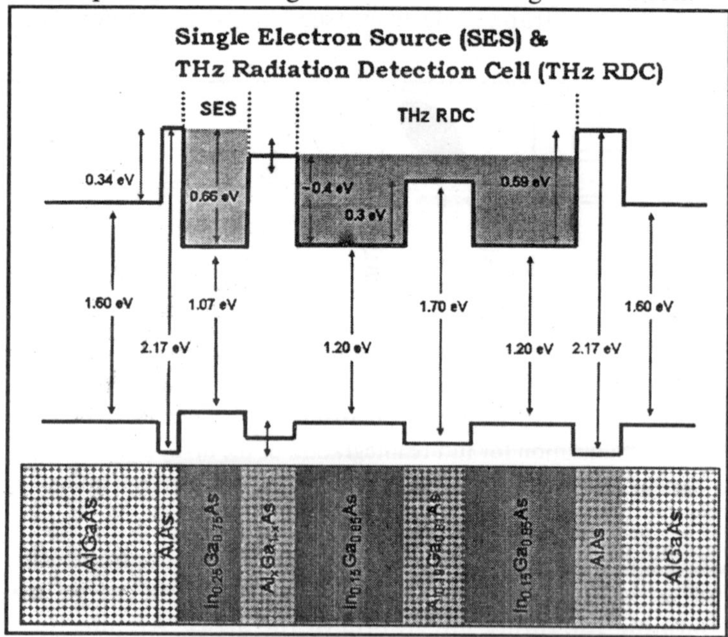

Fig. Vertically-grown Multi-layered Q-Dot Detector with Nanoscale Aperture.

Hence, the first required element is a QCD that responds to single-photon excitation with a change of electron polarization. In a practical implementation, both vertically-defined heterostructure-semiconductor Q-dots and laterally-defined nano-probe induced Q-dots are realistic options as both have merits for realizing the needed QCD component.

However, the fundamental principles of the detection methodology are perhaps best understood from the QCD which is illustrated in figure which consists of a vertically-grown multi-layered Q-dot set into a shielded configuration to define the nanoscale aperture.

Here, as shown in figure. the sensor consists of a three-well Q-dot system where the first-well serves as a single-electron source (SES) that allows for controlled injection of a single-electron into a double-well pair that makes up the THz Radiation Detection Cell (THz RDC).

This type of architecture is important for defining the proper quantum-states and the associated photonic processes that are necessary for enabling a statistically-based measurement of the incident radiation intensity that will actually allow for the THz microscopy function as is explained in the next subsection.

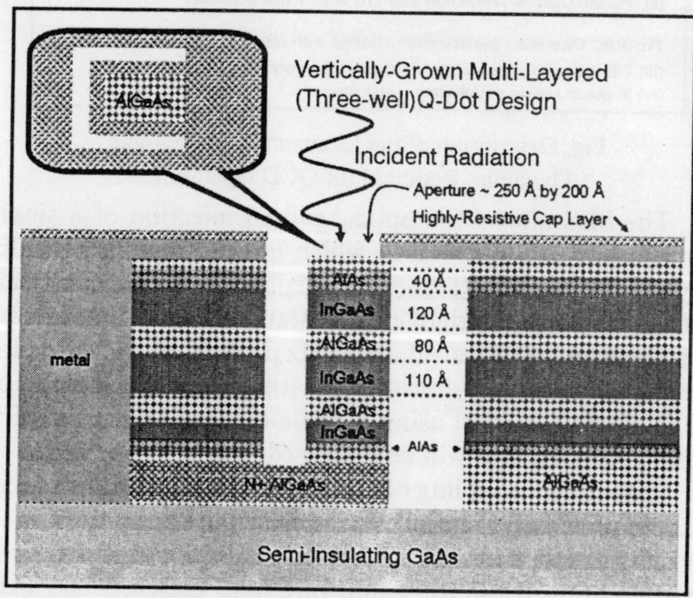

Fig. Semiconductor Heterostructure and Band Structure for Q-Dot Detector

QCD MEASUREMENT APPROACH

The QCD sensor utilizes a triple-well Q-dot system to enable monitoring of the induced polarizations of a single electron due to exposure to THz radiation.

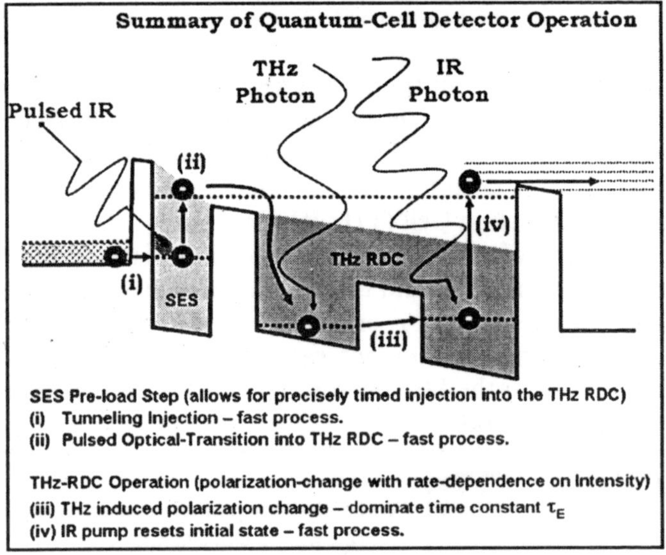

Fig. Description of the Electronic and Photonic Dynamics Required for QCD Operation.

The SES allows for optical-pulsed injection of a single electron into the THz-RDC, which has been asymmetrically engineered to possess THz energy-separations between the left ground-state and the right excited-state. Here, the microscopy measurement of the transmitted THz power through the nano-sized aperture is now amenable to statistical RF sampling of the RDC polarization using a single-electron-transistor SET based approach that will be discussed later.

Indeed, RF sampling of the resulting polarization process (i.e., via a SET impedance measurement) over many cycles of the combined THz-IR excitation enables a statistical determination of the transition rate (and therefore the THz field intensity) by the method as illustrated in figure. Here, the statistical rate of removal of the injected electron from the

RDC is dictated by the THz induced transition rate τE and since this is fundamental quantum mechanical process it can be assessed by sampling over many transitions to numerically determine the value of the incident field intensity.

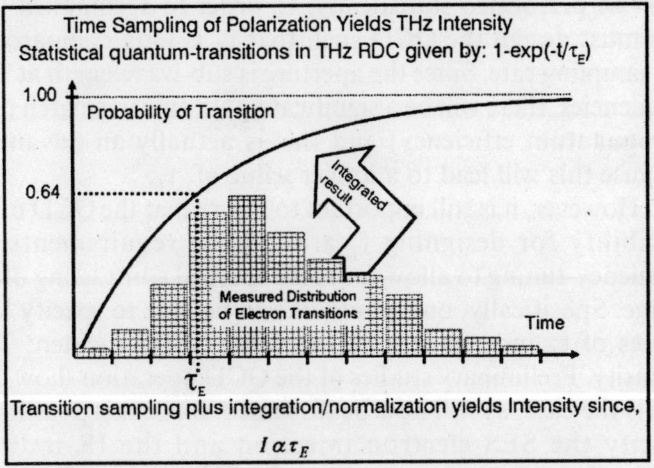

Fig. Statistical Determination of Transition Rate and Associated Intensity of Incident Electric Field.

QCD FEASIBILITY ANALYSIS

The approach of statistically counting the quantum transitions (which in a sense is counting the THz photons) to infer the value of the incident field intensity, *I*, at THz frequencies sets specific requirements on the electron coherence time and the energystate separation tuning ability associated with the QCD device. Specifically, the THz RDC must be physically compatible with state-of-the-art pulsed RF measurement techniques as the expected sampling rates (e.g., ~ μ secs) will dictate the ultimate sensitivity of the approach. This fact was illustrated in figures and since one needs to numerically integrate the results of the measured transition-distribution, a time increment equal to μ secs must be small in comparison to the value of τ_E.

Fortunately the THz QCD offers much latitude for engineering both the THz detection-frequency, f_{THz}, and the transition-rate $t_{E,}$ as will now be demonstrated. The technique presented for collecting a nanoscale-size pixel of information at THz frequencies relies on the technological capability for estimating the transition-rate of the excited electron which must be performed statistically. In order to accomplish this, one must design the QCD such that t_E is long compared to the sampling rate. Since the aperture is sub-wavelength at THz frequencies, there will be a significant photonic mismatch (e.g., low quantum efficiency) and this is actually an advantage because this will lead to a longer value of τ_E.

However, it is still important to insure that the QCD offers flexibility for designing t_E around the requirements for frequency-tuning to allow for some spectral contrasting of the image. Specifically, one would like to be able to specify long values of t_E independent of the value of the incident THz intensity. Preliminary studies of the QCD operation show that this is the case, as will now be discussed. Since it is possible to specify the SES electron injection and the IR-induced elimination of the electron after the THz excitation as very fast processes, an analysis of the THz RDC alone is sufficient to estimate the latitude in designing for long τ_E times.

As summarized in figure, a Schrodinger-equation based analysis of the RDC was applied to determine the ground and excited state of the trapped electron, and these results along with the associated spatial distribution for a zero-biased structure that yields f_{THz} = 600 GHz are illustrated. This same approach was then used to assess the both the ability to voltage-tune f_{THz} and to prescribe the dependency of τ_E on radiation intensity according to the RDC design and these results are summarized in figure.

Hence, the collective simulation results given in figures demonstrate that the QCD approach is viable for deriving the intensity (and power) of the transmitted THz beam over a nanoscale aperture, if a technological approach can be implemented for sampling the single-electron transitions at μ sec sampling rates. It is also noteworthy to point out that this ability to design for very long transition-times (or equivalently the ability to suppress the scattering rates to ~ 10^5 s^{-1}) has important ramifications for designing quantum gates for quantum computing applications as has been discussed in a companion publication.

Schrodinger Based simulations of THz RDC Operation

Numerical Simulations for Electron Wavefunction over 300-900 GHz Range

$$H_0\Psi = E\Psi \qquad H_0 = -\frac{\hbar^2}{2m^*}\nabla^2 + V(\vec{r})$$

$$\Psi(\vec{r}) = \frac{2}{\sqrt{L_x L_y}}\sin\left(\frac{2\pi x}{L_x}\right)\sin\left(\frac{2\pi y}{L_y}\right)\varphi(z)$$

$$\left(-\frac{\hbar^2}{2m^*}\frac{d^2}{dx^2} + V(z)\right)\varphi(z) = E_z\varphi(z)$$

$$E_z = E - \frac{\hbar^2}{2m^* L_x^2} - \frac{\hbar^2}{2m^* L_y^2}$$

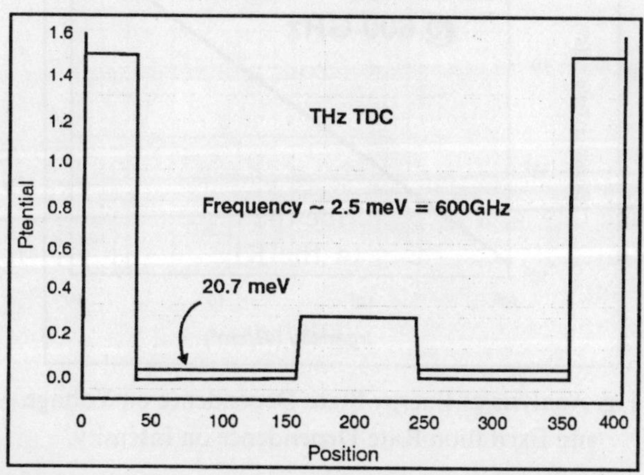

Analysis of Voltage Tuning & Transition-Time(τ_E)

THz RDC operation dominated by the THz-induced τ_E which is derivable as

$$w_{fi} = \frac{2\pi}{\hbar}\left|< \Psi_f \mid H'_{op}\Psi_i >\right|^2 \delta\left(E_f - \mathrm{E_i} - \hbar\omega\right)$$

$$< \Psi_f \left|H'_{op}\right| \Psi_i > = -\frac{eA_o}{2i}\left(E_i - E_f\right)\hat{\varepsilon} < \Psi_f \left|\vec{r}\right| \Psi_i >$$

$$A_0^2 = \frac{2I}{c\varepsilon_0\varepsilon_r\omega_{0p}^2}$$

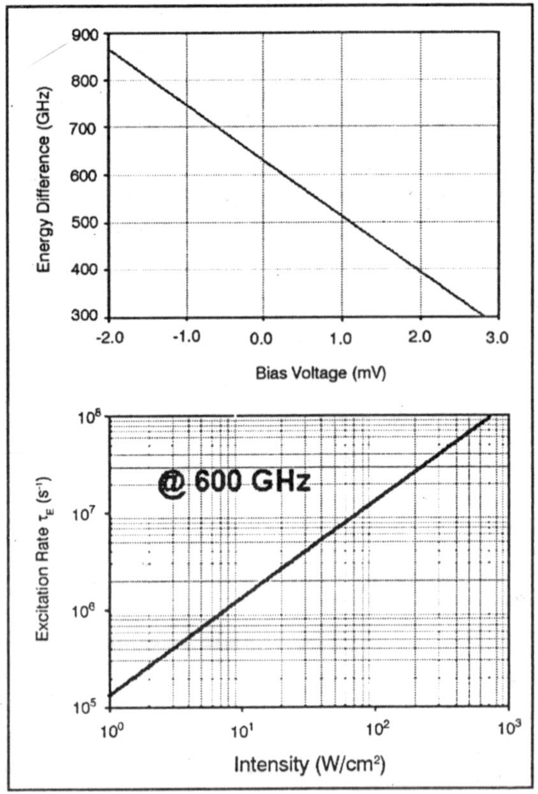

Fig. Analysis of Energy State Dependence on Voltage and Excitation Rate Dependence on Intensity.

SET-BASED READOUT CIRCUITRY DESIGN OVERVIEW

The previous scheme for achieving a THz microscopy

capability has been shown to be feasible if a technological approach can be implemented to sample and readout the single-electron transitions.

Fortunately, recent advances make it possible to utilize singleelectron- transistors (SETs) to measure the charge state within a Q-dot with sub-µsecond temporal resolution6 so the real challenge is to engineer QCD-compatible RF-SET readout circuitry to enable frequency-domain-multiplexing measurements of the polarization state.

While the discussion given earlier presented only one compound semiconductor material system design for a vertical QCD, it is possible to prescribe the required type of QCD device (i.e., with three quantum-wells) using at least two different semiconductor heterostructure systems as illustrated in the bottom of figure.

These material systems allow for the needed latitude in defining the well-widths for the SES section and the THz-RDC section, as these factors are very important for achieving the goals of the THz-QDC and the statistical sampling scheme. For example, the material compositions of the second barrier from-the-right in figure (i.e., AlxGa1-xAs upper material system and AlxIn1-xAs lower material system) allow for adjustments in defining the sub-band energies for electron injections into the SES and transfers into the THz- RDC. In addition, it is necessary to define the SES section to allow for rapid and controllable injection of a single-electron into the THz-RDC.

Also, a strategy must be devised for rapidly removing the electron after the THz photon has been detected via the transition of the single-electron from the left-well to the right-well of the THz-RDC. Figure illustrates two variations in the QCD design that allows for prescribing these processes so that the transition-rate, t_E, will be the dominant time mechanism. As shown, it is possible to utilize pulsed IR excitation to precisely define the time when the singleelectron is injected into the ground-state of the THz-RDC, and this single-electron can be removed after the THz-excitation to the excited-state by assisted IR pumping.

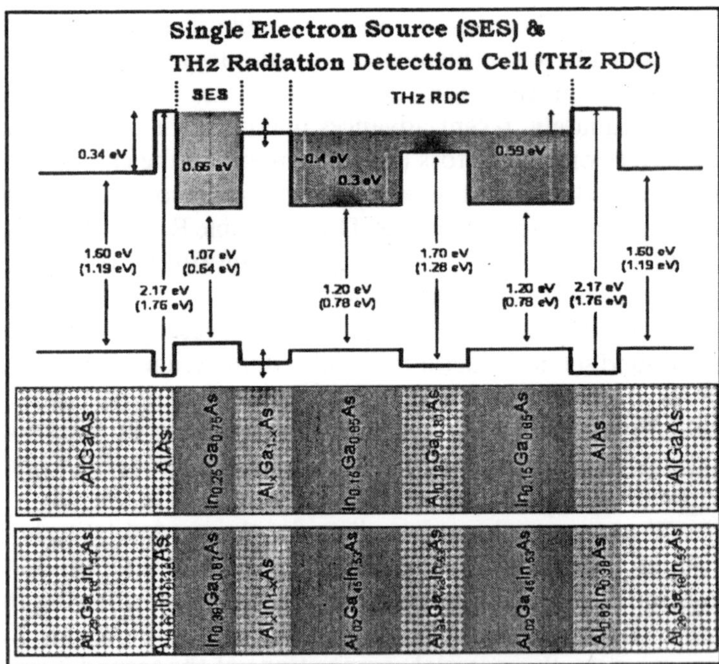

Fig. Two Material Designs for the Vertical QCD

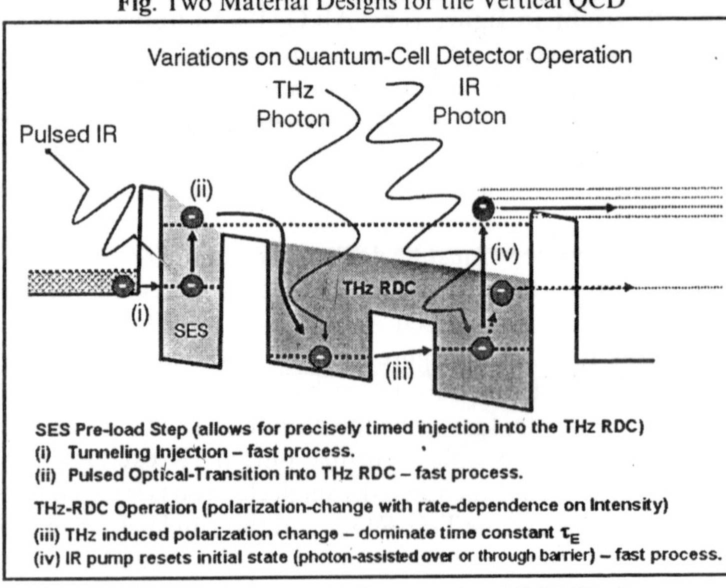

Fig.Two Variations in IR Pump-assisted QCD Design.

Here two important variations are possible with the single-electron being raised above the rightmost barrier, and possibly the more attractive approach of only raising the single-electron to a sub-band state that has been made amenable to tunneling through the right-most barrier (see modified barrier-structure in green and associated process given by the blue-dashed arrows). The ability to obtain single-photon counting using methods such as avalanche photon detectors and negative electron affinity photocathode photomultipliers has thus far been limited to the visible and infrared regions.

The vertical QCD which utilizes a triple-well quantum (Q) dot system of the type illustrated in figure offers a novel approach to sense THz radiation. Here, the detector is first primed into active-mode by tunnel injection into the top Q-dot (QD1) of the SES followed by an IR pulse that puts the electron into the middle Q-dot (QD2) of the THz-RDC. This electron will remain in the QD2 until a THz photon induces the electron's transition to QD3. Finally, an IR photon ejects the electron from QD3 thus resetting the detector. Since the electron injection into the QCD system and ejection from the detector are quick transitions, only the middle Q-dot (QD2) will be occupied for a significant period of time. Consequently, to successfully read-out the state of our triple Q-dot system one must be able to differentiate between the two possible states:

- If THz photons are present, the electron will quickly be ejected from the entire QCD system and no electrons will be present in QD2.
- If no THz photon is present, QD2 will remain occupied by an electron.

The approach proposed here for distinguishing between these two states is to utilize a carbon nanotube (CNT) RF single-electron transistor (RF-SET) that is capacitively coupled to the quantum-dot based detector as shown in figure. An RF-SET is advantageous for this application because it is an ultra-sensitive electrometer capable of differentiating the presence of single charges and it readily allows for multiplexing where by multiple RF-SETs can be readout simultaneously with each identifiable by its characteristic resonant frequency. Researchers have demonstrated RF-SET charge of sensitivity of order $10^{-5}\, e/\sqrt{Hz}$.

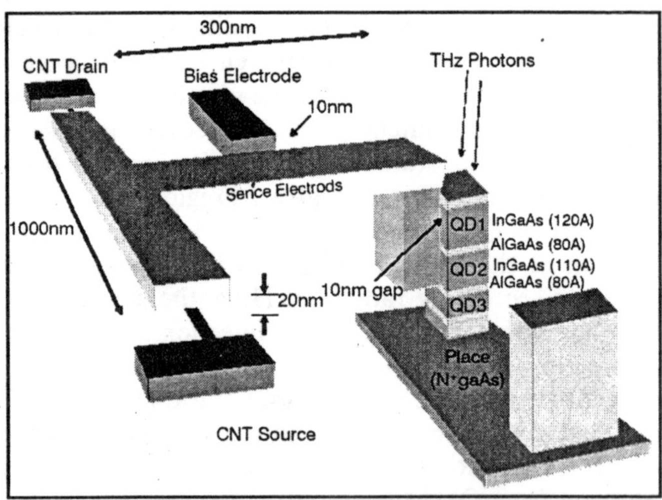

Fig. The proposed structure of the SET-based readout circuit for THz detector. The quantum detector cell (QDC), which is the triple Q-dot system, is capacitively coupled through the sense electrode to a carbon based nanotubes RF-SET. A bias-electrode is used to control the bias potential on the sense electrode.

A recent advance at Yale has allowed this to be achieved at rates with sub-microsecond temporal resolution with the so-called RF-SET, and this has been used by Rimberg to measure the charge state of a quantum dot in real time.

Therefore, based on this preliminary work by other groups, it seems very feasible to determine the charge state of a quantum dot using a single electron transistor. Furthermore, utilizing a carbon nanotube as our RF-SET affords us the advantage of having a higher operating temperature as well as being easier to fabricate compared to its GaAs counterpart. Due to fabrication constraints, the QDC and the RF-SET can not be fabricated directly adjacent to one another. To reconcile this problem, a sense electrode is introduced to act as the intermediary by capacitively coupling both the triple quantum-dot system as well as the CNT RF-SET to one another. In other words, when the RDC is in the primed-state (i.e. an electron is residing in QD2) a voltage will be induced on the sense electrode.

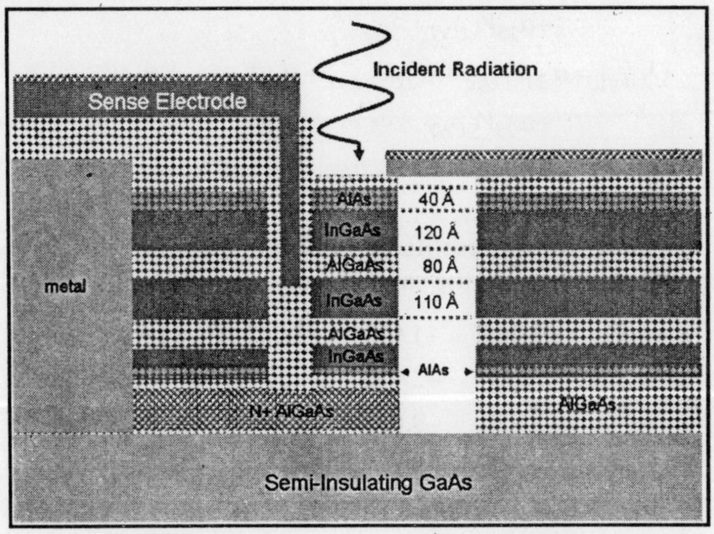

Fig. A Side View of the Sense Electrode for RF-SET Integrated with the QCD Device.

Since this electrode also doubles as the gate for the RF-SET, the induced voltage change will be translated over to the carbon nanotube RFSET in the form of a change in the nanotube's impedance. A bias electrode is present to properly adjust the sense electrode's potential so to maximize the change in RF-SET impedance between the two states, $Z_{CNT\text{-}RFSET}$.

SET-READOUT CIRCUITRY SIMULATIONS

To differentiate between the two possible QDC states, i.e. whether QD2 is occupied or not, a reliable value for the capacitive coupling between the quantum dots and the sense electrode is needed. To that end, simulations for the capacitance of the proposed structure shown in figure were performed using the software programme Fast Model. For simplicity purposes the bias electrode was not included in these simulations because it will have a negligible affect on the results. The relation between the charge and voltage vectors can be written out as a linear set of equations as

$$Q_{1;elec} = \alpha_{11}V_{1;elec} + \alpha_{12}V_{2;dot1} + \alpha_{13}V_{3;dot2} + \alpha_{14}V_{4;dot3} + \alpha_{15}V_{5;CNT}$$

$$Q_{2;dot1} = \alpha_{21}V_{1;elec} + \alpha_{22}V_{2;dot1} + \alpha_{23}V_{3;dot2} + \alpha_{24}V_{4;dot3} + \alpha_{25}V_{5;CNT}$$

...

...

where α_{ii} represents the coupling capacitance. The simulated results are the capacitance matrix

$$\alpha = \begin{pmatrix} 97.34 & -3.762 & -1.934 & -1.757 & -39.53 \\ -3.762 & 8.135 & -2.930 & -0.2794 & -0.1004 \\ -1.934 & -2.930 & -9.022 & -3.176 & -0.0890 \\ -1.757 & -0.2793 & -3.107 & 7.358 & -0.1771 \\ -39.53 & -0.1004 & -0.0890 & -0.1771 & 51.45 \end{pmatrix} \begin{matrix} elec(aF) \\ dot1 \\ dot2 \\ dot3 \\ CNT \end{matrix}$$

To convert the capacitance matrix values a into physically meaningful capacitance values, represented by the traditional variable *C*, consider the circuit shown in figure. For simplicity, only the interconnecting capacitances associated with the top quantum dot (QD1) have been shown.

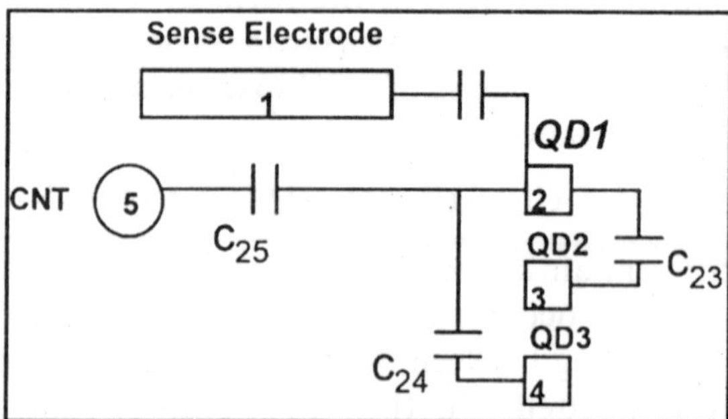

Fig. An example showing all the capacitances of the top quantum dot (QD1) interconnecting with the other structures of the QCD device and SET readout circuit elements.

Here we can represent $Q_{2;dot1}$ as

$$Q_{2;dot1} = C_{21}\left(V_{2;dot1} - V_{1;elec}\right) + C_{22}\left(V_{2;dot1}\right)C_{23}\left(V_{2;dot1} - V_{3;dot2}\right)$$
$$+C_{24}\left(V_{2;dot1} - V_{4;dot3}\right) + C_{25}\left(C_{2;dot1} - V_{5;CNT}\right)$$
$$Q_{2;dot1} = V_{1;elec}\left(-C_{21}\right) + V_{2;dot1}\left(C_{21}C_{22} + C_{23} + C_{24} + C_{25}\right)V_{3;dot2}$$
$$\left(-C_{23}\right) + V_{4;dot3}\left(-C_{24}\right) + V_{5;CNT}\left(-C_{25}\right)$$

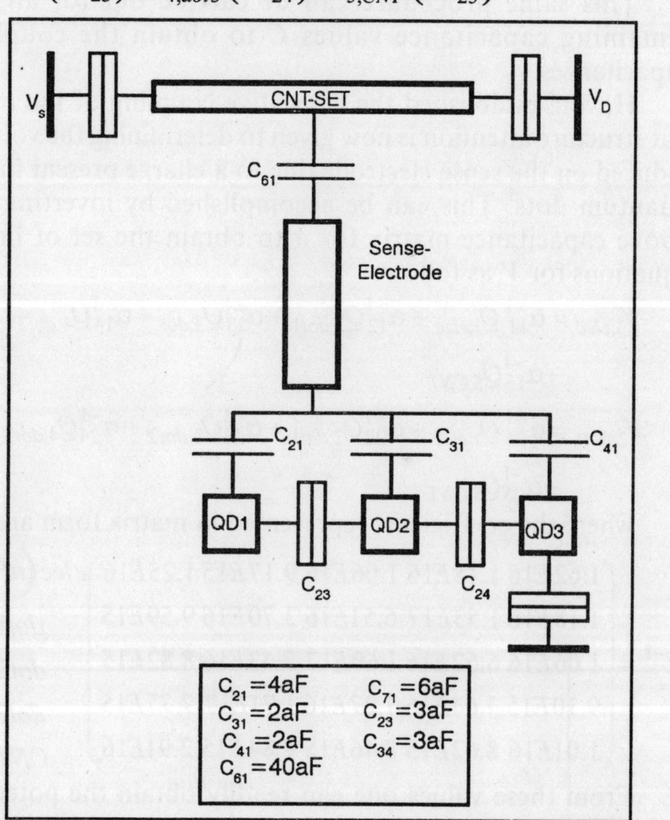

Fig. A schematic of the proposed read-out circuit for the THz detector showing only the capacitances contacting the sense electrode. The capacitance values given were obtained from the simulation study results.

and by matching coefficients of $Q_{2;dot1}$ in (1) with (2) one obtains

$$\alpha_{21} = -C_{21}$$
$$\alpha_{22} = C_{21} + C_{22} + C_{23} + C_{24} + C_{25}$$
$$\alpha_{23} = -C_{23}$$
$$\alpha_{24} = -C_{24}$$
$$\alpha_{25} = -C_{25}$$

This same procedure can be carried out for all the remaining capacitance values C to obtain the coupling capacitances.

Having established the capacitive coupling of the read-out structure attention is now given to determining the voltage induced on the sense electrode due to a charge present in the quantum dots. This can be accomplished by inverting the above capacitance matrix for a to obtain the set of linear equations for V as follows

$$V_{1;elec} = \alpha_{11}^{-1} Q_{1;elec} + \alpha_{12}^{-1} Q_{2;dot1} + \alpha_{13}^{-1} Q_{3;dot} + \alpha_{14}^{-1} Q_{4;dot3} + \alpha_{15}^{-1} Q_{5;CNT}$$

$$V_{2;dot1} = \alpha_{21}^{-1} Q_{1;elec} + \alpha_{22}^{-1} Q_{2;dot1} + \alpha_{23}^{-1} Q_{3;dot2} + \alpha_{24}^{-1} Q_{4;dot3} + \alpha_{25}^{-1} Q_{5;CNT}$$

where the coefficients represented in matrix form are

$$\alpha^{-1} = \begin{pmatrix} 1.62E16 & 1.18E16 & 1.06E16 & 9.17E15 & 1.25E16 \\ 1.18E16 & 1.53E17 & 6.51E16 & 3.70E16 & 9.59E15 \\ 1.06E16 & 6.52E16 & 1.60E17 & 7.43E16 & 8.82E15 \\ 9.10E15 & 3.64E16 & 7.27E16 & 1.71E17 & 7.77E15 \\ 1.01E16 & 8.62E15 & 8.46E15 & 7.84E15 & 2.91E16 \end{pmatrix} \begin{matrix} elec\left(aF^{-1}\right) \\ dot1 \\ dot2 \\ dot3 \\ CNT \end{matrix}$$

From these values one can readily obtain the potential induced on the sense electrode for our two cases, i.e. (1) when an QDC is in the primed state with an electron resides in QD2, or (2) when a THz photon excites the QD2 electron causing it to be ejected from the QDC.

The corresponding voltage that would exist on the sense electrode if QD2 had a charge $Q_{3,dot\ 2} = e$ with all other structures having $Q = 0$ is,

$$V_{1;elec}\left(Q_{2;dot1}=n\cdot e\right)=\alpha_{11}^{-1}Q_{1;elec}+\alpha_{12}^{-1}Q_{2;dot1}+\alpha_{13}^{-1}Q_{3;dot2}$$

$$\alpha_{14}^{-1}Q_{4;dot}+\alpha_{15}^{-1}Q_{5;CNT}$$

$$=0+0+(1.06E16)(n\cdot e)+0+0$$

$$=1700\mu V.n$$

For case (2), when the electron is ejected from the QDC the potential on the sense electrode is 0 V.

This yields a DV = 1.7 mV between the two cases. At this point, it is possible to determine how much the impedance of the SET will change as a result of a change in the gate voltage of ΔV = 1.7 mV. The capacitance between the CNT-SET and sense electrode has been approximated to be C51 = 40 aF which means the distance between conductance peaks of the SET would be $\Delta V_g = e/C_{51} = 4\ mV$. Consequently, it is possible to derive the useful approximation,

$$\Delta G=\frac{\partial G}{\partial V_g}\Delta V_g$$

$$\approx\frac{G_{\max}}{2mV}\cdot 1700\mu V$$

$$\approx 0.85G_{\max}$$

If one assumes G_{max} = 10 μ S (corresponding to R = 100 kΩ) and if the RF-SET is biased such that one of the double quantum-dot states gives a SET-conductance of $G_1{=}G_{\max}$ then the other state would be $G_2=G_{\max}-\Delta G=1.5\mu S$. In terms of the impedance seen by the matching circuit this would be comparable to distinguishing 100 kΩ from a 670 kΩ resistor. This is entirely feasible for achieving a successful readout with the available high signal-to-noise as will be discussed next in the impedance-matching section.

SET-READOUT TANK CIRCUIT DESIGN

Fundamental to the RF-SET readout circuitry is the accompanying LC tank circuit which will transform the relatively high impedance of the carbon nanotubes RF-SET down to the characteristic impedance of 50 W. In addition, when multiple THz detectors are integrated together, multiplexing of their signals can be accomplished by choosing unique LC tank-circuit resonant frequencies for each THz detector cell. The tank circuit shown in figure consists of the parasitic capacitance to ground and an inductor ($C_{parasitic}$ is typically on the order of a fraction of a pF). Although the actual parasitic capacitance will vary between samples, for illustrative purposes we will assume $C_{parasitic}$ = 0.18 pF and a L_{tank} = 710 nH chip inductor with a resistance at resonance of $R_{L\text{-}Tank}$ = 10 W. For this illustrative case, the S11 results have been simulated using the software ADS and as the RF-SET resistance changes from $R_{RF\text{-}SET}$ = 100 kW to 130 kW the S11 will change by over

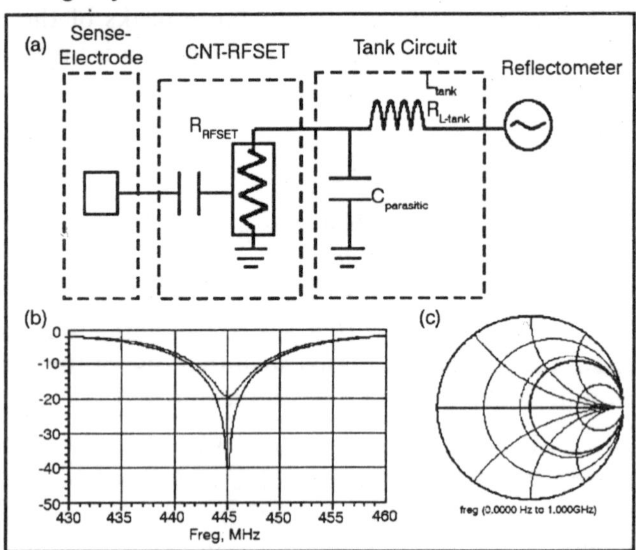

Fig. (a) A schematic of the LC-tank circuit used to transform the high impedance of the RF-SET down to the characteristic impedance of Z = 50 W. (b) Using the simulation software ADS, S11 is plotted for the case when $R_{RF\text{-}SET}$ = 100 k and when it is 130 kΩ along with the corresponding Smith chart, (c). The simulated component values of the tank circuit were $C_{parasitic}$ = 0.18 pF, L_{tank} = 710 nH, and the resistance of the inductor at its resonant frequency $R_{L\text{-}Tank}$ = 10 Ω.

In practice, though one should expect a lower DS11 value

due to the difficulty of precisely matching the RF CNT impedance to 50 W due in part to: (1) the constraint of limited bandwidth at which the cryogenic RF amplifiers maintain optimal bandwidth, (2) the limited availability of specific valued chip-inductors, (3) variation in the parasitic capacitance between samples. Preliminary works has also been conducted to build and test an RF tank circuit with L_{tank} = 10 nH and $C_{parasitic}$ ~ 7 pF and to demonstrate it at room temperature on chip resistors of various values, from 100 W to 100 kΩ. The results are given in figure. This demonstration circuit shows the ability to match to a 100 kW load and confirms the feasibility of the SET-based readout approach.

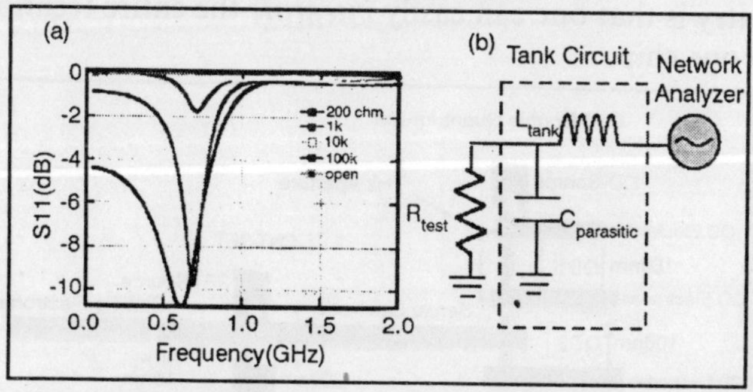

Fig. A RF tank circuit (as illustrated in (b)) with L_{tank} = 10 nH and $C_{parasitic}$ ~7 pF was tested to demonstrate how an impedance matching circuit would transform various resistance values (as given in (a)).

AN ALTERNATIVE ALL-NANOTUBE INTEGRATED DESIGN CONCEPT

While the proposed concept design for the THz detector is feasible, it is also possible to suggest alternative designs that might offer advantages. As a potentially viable alternative to designing a vertically-defined heterostructure-semiconductor Q-dot system (e.g., InGaAs/AlGaAs triple Q-dot detector with optical injection), one could conceivably obtain the same result using a laterally-defined single-carbon-nanotube with multiple tops-gates as shown in figure.

Compared to traditionally designed quantum dots (lateral gating of a 2DEG, vertical pillars), nanotube dots are by far the smallest and will allow for the simplest route to nanoscale spatial resolution. In addition, lateral gating of 2DEGs requires many electrodes to define each dot whereas a nanotube needs significantly fewer electrodes. Since the nanotube diameter is so small, transverse energy levels are "frozen out", and the material is inherently one dimensional, simplifying the quantum excitation spectrum considerably. In addition, the nanotube small size causes the excitation energy scales in nanotube dots to fall naturally in the THz regime, whereas in lateral dots and the vertical design, they can be typically 10-100 GHz. Finally, a very important advantage of using nanotubes to define both the detection Q-dots and the readout circuitry is that one can easily *integrate* the entire technology onto one chip.

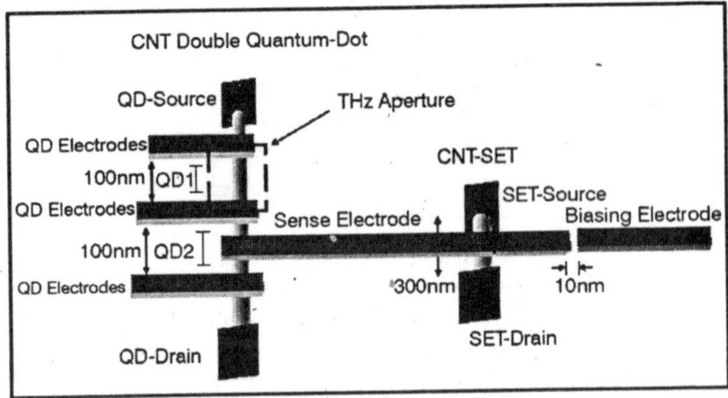

Fig. This alternative structure for the THz detector and read-out circuit functions in the same manner as the one defined in figure. 10 but utilizes QDs that are formed by nanotubes and potential electrodes. This structure offers fabrication advantages because only nanotubes and metallic gates would be involved.

CONCLUSIONS AND FUTURE GOALS

This paper has presented a novel nanoscale-engineering methodology along with descriptive nanotechnology devices needed for the first-time development of a microscope-system capable of collecting terahertz (THz) frequency spectroscopic signatures from microscopic biological (bio) structures. The proposed THz microscope will represent a first of its kind (i.e., ability to perform THz spectral interrogation of microscopic targets) and it is proposed to be built upon completely new nanoelectronic concepts. The specific technological advantages include:

- Novel QCD design that allows for sampling the influence of THz-frequency photons on single-electron polarization which allows one to infer the intensity of the THz radiation over subwavelength apertures;
- Revolutionary CNT RF-SET technology that allows for ultrafast sampling of single-electron motion and the rapid readout of the information associated with a nanoscale THz sensing pixel; and when these technologies are implemented into a THz microscopy system they should allow for a unique and unprecedented long-wavelength spectral imaging capability. The envisioned technology targets the goal of achieving the type of microscopy illustrated in the scenario of figure. Specifically, a scanning pixel array (or minimally a single pixel probe) is desired that will be effective for shadow-imaging biological targets with pixel resolution that is small in comparison to the structural features of the object.

Here, the target object is assumed to be illuminated from below by THz radiation that is variable both in intensity and signal frequency. As hypothetically illustrated, expected bio-targets might be single cells or viruses of micron-size and below, so it will be necessary to define sensor elements capable of measuring the transmission of THz frequency photons through materials/agents that possess geometrical structure on the nanoscale. Therefore, the aperture of the sensor must be on the order of 10-100's of nanometers while still being able to sense across a reasonable THz bandwidth with very high detection sensitivity. This type of THz microscopy capability is attractive to enable the contrasting of structural features within the target that might exhibit a strong dependence on spectral absorption as illustrated in the top of figure.

For example, the circular green sub-unit within the cell

(see optical visual image on the far left) could be transparent at THz frequency F_1, but demonstrate a condition of opacity at THz frequency F_2, and it possesses an inner light-green figure-eight feature that might absorb strongly only at a third THz frequency F_3. While the exact scenario depicted in figure is hypothetical, THz spectral signatures of the type assumed have been demonstrated in nearly perfect bulk-samples of bio-molecules and bio-cellular materials. Hence, a THz microscope of the type described would have profound implications for scientific inquiry and such a sensing capability would be valuable for practical applications related to medical diagnostics and bio-threat agent defence and security. Future work will focus on the further investigation and refinement of the technology presented here, and the pursuit of initial technology demonstrations will be conducted through a teamed multidisciplinary effort.

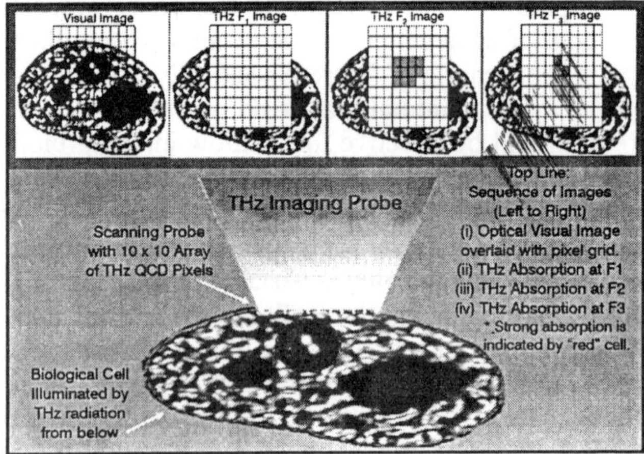

Fig. Illustration of the Operation Scenario of the Suggested THz-Frequency Transmission Microscope.

Chapter 16

Optical Pumping

Sometimes you want to magnetize a gas, to align all of the little magnetic moments of the gas atoms in the same direction.

You might plan to have somebody inhale the polarized gas, then look at the insides of their lungs using magnetic-resonance imaging, or you might want to investigate the electronic level structure of the gas atoms, and polarizing them is the first step in a spectroscopy programme.

There are lots of times, in a modern research lab, where you want a polarized gas, and the usual way to produce it is by optical pumping.

In this lab you will learn basic optical pumping, and you will practice it on a gas of Rubidium atoms. Rubidium is a fairly simple atom. Its electronic structure closely approximates that of a Hydrogen atom, so we can get a pretty good theoretical understanding. However, even in Hydrogen-like atoms there are interesting effects that can be investigated spectroscopically using optical pumping, and we will do that in this lab.

RUBIDIUM'S ATOMIC ENERGY LEVEL STRUCTURE

Rubidium in its atomic state has just one valence electron and can be well approximated by a one-electron-atom model. Its nuclear properties are different from Hydrogen, however, and this will give it a different energy-level structure. There are two commonly-occurring isotopes of Rubidium in nature, ^{85}Rb, with a natural abundance of 72%, and ^{87}Rb, with an abundance of 28%. In this section you will learn all about ^{87}Rb, but the structure of ^{85}Rb follows the same general priciples.

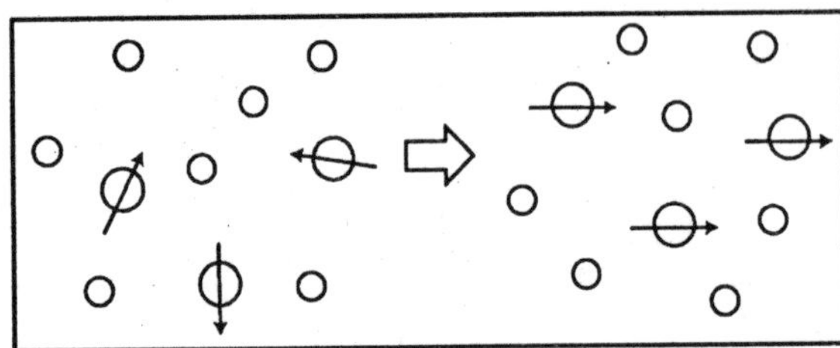

Fig: Optical pumping can be used to polarize a gas of atoms that have magnetic dipole moments. In practice, these atoms are often mixed with a nonpolar buffer gas, which helps keep the polarized atoms from touching with the walls of the container and losing their polarization.

S=1/2
I=5/2
^{85}Rb (72%)
S=1/2
I=3/2
^{87}Rb (28%)

Fig: There are two commonly-occurring isotopes of Rubidium found in nature, ^{85}Rb and ^{87}Rb. Both have only one valence electron and can be approximated as one-electron atoms. The major difference between the isotopes is in the nuclear spin I.

E=0 (ionized)
etc...
4s 4p 4d 4f
3s 3p 3d
2s 2p
1s
e
p
E
l

Fig: The Basic Energy-level Structure of a One Electron Atom.

Bound states have negative energy and are quantized, with different allowed amounts of angular momentum for each energy level. To first approximation, all of the different angular momentum states have the same energy for a given energy number n. A crude model of a one-electron atom that neglects both the spin of the electron and the nucleus gives a level structure like that shown in figure.

Here, the levels are labeled by their energy quantum number n and the orbital angular momentum ‘, and $\ell = s, p, d, f, ...$ the notation is n‘, where n = 1, 2,... and. The use of letters to represent ‘ is left over from the old days of spectroscopy, before people knew what was going on inside atoms, and the letters stand for “sharp,” “principal,” “diffuse,” and “fundamental.” (After f they go as the alphabet: *g, h, i,* etc.) This notation is known, appropriately, as spectroscopic notation.

The spinless model does a pretty good job, considering how simple it is. A low-resolution spectrometer will see an emission or absorption spectrum of a one-electron atom (any alkali atom) that really does agree with the predictions of this model.

However, if we use a higher-resolution instrument, like the optical-pumping apparatus in this lab, additional structure will show up that the spinless model cannot account for. There are many corrections to the basic, spinless model of the atom, but we will only be concerned with three: taking into account the spin of the electron and of the nucleus, and the addition of an external magnetic field. The spin of the electron is important because it couples with the orbital angular momentum.

Essentially, the electron feels an effective magnetic field as it moves through the electrostatic field of the nucleus. Since the electron has an intrinsic magnetic moment, due to its spin, its energy level will be higher if it is aligned opposite this effective magnetic field than if it is aligned with it. This leads to an additional term in the Hamiltonian of the form L·S, which changes the eigenstates ever so slightly. Only the states with non-zero orbital angular momentum are affected, and the net result is that they are split into multiple levels, depending on the orientation of the spin. This is known as fine-structure splitting and is illustrated in figure.

Once we take electron spin into account, we have to modify our spectroscopic notation. The new notation is called, appropriately enough, the modified spectroscopic notation or Russel-Saunders notation, and it goes like this. Each state is labeled by $^{2S+1}L_J$, where S is the electron spin, L is the orbital angular momentum, and J = L + S is the total angular momentum. The energy quantum number n is dropped. For our single-electron atom, S is always 1/2, but the notation allows treatment of multi-electron atoms with their spins added in proper quantum fashion. The orbital angular momentum is again labeled with the old spectroscopic letters (S, P, D, etc.), but this time they are capitalized. The quantum number J ranges from |L – S|

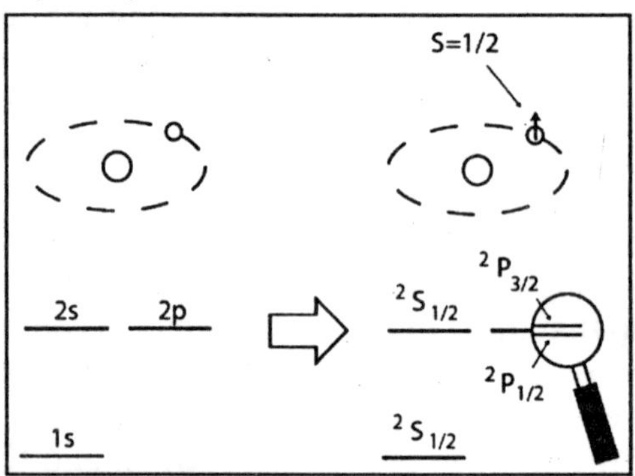

Fig. Fine Structure– Taking the Spin of the Electron into Account Leads to a Small Splitting of the Levels with Orbital Angular Momentum.

Now that there are two angular momenta to consider, orbital and spin, we must modify our notation. The new notation is $^{2S+1}L_{J,}$ where S is the electron spin (1/2 for one electron), L is the orbital angular momentum (S,P,D,etc.), and J = L+S is the total angular momentum. The energy quantum number n is dropped. to |L + S|, according to the rules of addition of angular momentum. Adding the spin of the nucleus I introduces three additional terms to the Hamiltonian, each of which are related to the nuclear magnetic moment associated with I.

You can look up a detailed treatment of this in your

favourite quantum mechanics text, but the end result is that the electron's orbital angular momentum, the electron's spin, and the nucleus' spin all add up to a total angular momentum F = I +J, with eigenvalues |I – J| < F < |I + J|. Each of these eigenstates of F has a slightly different energy, due to the coupling between the electron and the nucleus' intrinsic magnetic field, as illustrated in figure. Finally, applying an external magnetic field further splits each of the F levels according to their projection number M. All this says is that the energy of the system (atom plus magnetic field) is different for

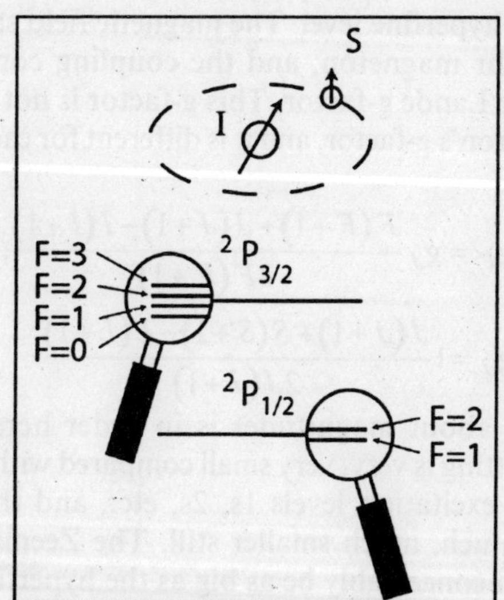

Fig. Hyperfine Structure– Adding the Spin of the Nucleus Further Splits the Levels.

Here we have a closeup of the 2p levels, split into $^2P_{1/2}$ and $^2P_{3/2}$ by the fine-structure correction. These fine-structure levels are split according to the total angular momentum F = I + J. This example illustrates the splitting for 87Rb, which has I = 3/2. The $^2S_{1/2}$ level also gets split, in the same way as $^2P_{1/2}$. different orientations of the atom, which is a perfectly reasonable assertion. This splitting is known as the Zeeman effect.

The fundamental physics of the Zeeman effect is essentially

the same as that of the fine and hyperfine splittings in that all three are due to coupling between magnetic fields and magnetic moments.

The difference is that the Zeeman effect is due to a magnetic field we apply in the lab, whereas the fine and hyperfine splittings are due to intrinsic magnetic fields that occur inside the atom itself. If the magnetic field is relatively weak, the Zeeman splitting is given by a simple expression.

$$E_z = g_F \mu_B BM$$

Here, E_Z is the Zeeman energy, the difference in energy between the state with projection number M and the unperturbed hyperfine level. The magnetic field strength is B, μ_B is the Bohr magneton, and the coupling constant gF is known as the Lande g-factor. This g-factor is not the same as the bare electron's g-factor, and it is different for each hyperfine level.

$$g_F = g_J \frac{F(F+1)+J(J+1)-I(I+1)}{2F(F+1)},$$

$$g_J = 1\frac{J(J+1)+S(S+1)-L(L+1)}{2J(J+1)}$$

A word about magnitudes is in order here. The fine-structure splitting is very, very small compared with the spacing between the excitation levels 1s, 2s, etc., and the hyperfine splitting is much, much smaller still. The Zeeman splitting, however, can conceivably be as big as the hyperfine splitting, or even bigger, if a strong enough magnetic field is applied. In practice, it is difficult to generate high enough fields to get well into the strong-field regime, but even in this lab we will begin to see deviations from the weak-field limit at our highest field values.

- Evaluate the Land′e g-factors for the highest-F hyperfine levels of the $^2S_{1/2}$ state, in both isotopes of Rubidium.
- Evaluate the Zeeman splitting for the highest-F hyperfine levels of the $^2S_{1/2}$ state, for both isotopes of Rubidium. What frequencies of electromagnetic radiation would you need to induce transitions

between these Zeeman levels in a 1 Gauss magnetic field?

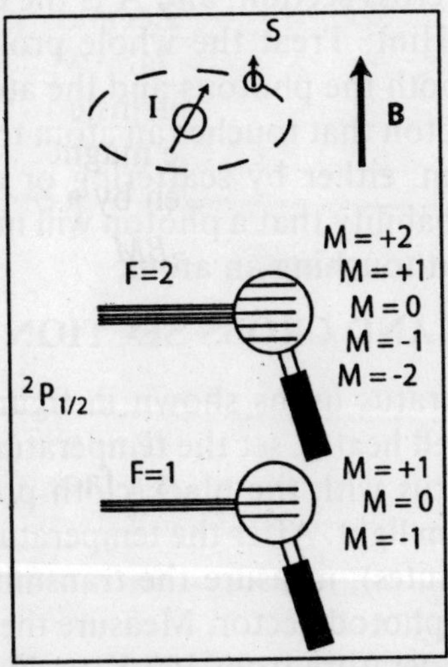

Fig. Zeeman Splitting– If we add an external magnetic field, each of the individual F levels splits according to its angular-momentum projection number M, i.e. the orientation of its magnetic moment. Of all the splittings we have considered so far, this is the only one whose magnitude can be comparable to the next larger correction, in this case the hyperfine splitting.

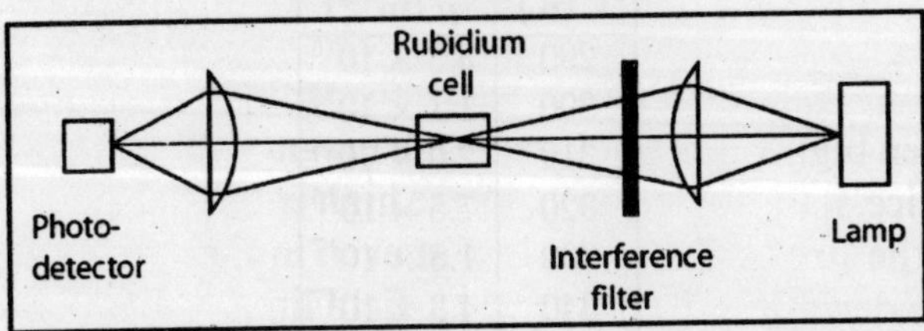

Fig. Setup for measuring absorption vs. temperature. Use unpolarized light for this part of the lab, with only the interference filter to block everything but the 795 nm line.

ABSORPTION AND CROSS SECTIONS

Show that the intensity of light falls off exponentially as the light travels through an absorbing medium. Specifically, show that $I = I_0 e^{-\sigma \rho x}$, where I is the intensity of the light after it travels a distance x in the medium, I_0 is the initial intensity, Γ is the effective cross section, and A is the density of atoms in the medium. Hint: Treat the whole problem classically, pretending that both the photons and the atoms are spheres. Consider any photon that touches an atom to be taken out of the incident beam, either by scattering or absorption, and calculate the probability that a photon will make it through a distance x without touching an atom.

ABSORPTION AND CROSS SECTION

Set the apparatus up as shown in figure 9. Locate the controls for the cell heater, set the temperature to 300 K, and cover the apparatus with the black cloth provided to block out any stray room light. After the temperature has stabilized (about thirty minutes), measure the transmitted intensity of the light with the photodetector. Measure the transmission as a function of temperature from 300 K to 400 K, then fit the theoretical absorption you calculated in the prelab to your data. You will need to know the density of the Rubidium gas as a function of temperature, which is given approximately in Table I. Table I: Density of Rubidium gas as a function of temperature.

T (K)	ρ (m^{-3})
290	3.3×10^{15}
300	1.1×10^{16}
310	2.9×10^{16}
320	7.5×10^{16}
330	1.8×10^{17}
340	4.3×10^{17}
350	8.3×10^{17}
360	1.5×10^{18}
370	3.7×10^{18}
380	6.3×10^{18}
390	1.2×10^{19}
400	2.4×10^{19}

(When performing the fit, don't forget to add a constant

intensity to your theoretical formula to account for nonresonant light that leaks into the system.) From your fit, calculate a value for the effective cross section Γ. How does this compare with the geometric cross section of Rubidium or the photons?

CIRCULARLY-POLARIZED LIGHT

OPTICAL PUMPING

Now consider what happens when a photon is introduced that has the right energy to lift an electron out of the 1s, $^2S_{1/2}$ state and put it into the 2p, $^2P_{1/2}$ state. The hyperfine levels of the 1s, $^2S_{1/2}$ state are close enough together that they are more or less equally populated.

That is, their spacing is much less than k_BT, and the valence electron in any given atom has about an equal probability of being in any F state with any M. An electron in any one of these states will get excited by the photon into one of the (F, M) states in $^2P_{1/2}$, but there are only certain states it can go to.

For example, the electron's final M value, after it lands in the excited state, can't differ from its initial M value by more than one. A photon is just not capable of changing M by more than one. Whether M changes by +1, –1, or 0 depends on the nature of the photon. If the applied magnetic field is parallel to the direction of propagation of the photon, then a right-circularly-polarized photon will always induce transitions that have

$$\Delta M = +1.$$

Left-circularly-polarized light produces

$$\Delta M = -1.$$

The same thing is true for emission. An electron can fall from the 2p, $^2P_{1/2}$ level into the 1s, $^2S_{1/2}$ level and emit a photon with right or left circular polarization, depending on whether ΔM is +1 or –1. For emission, both cases are equally likely. For absorption, we control which kind of photon is incident, so we can control ΔM. Now we are in a position to polarize atoms. If we shine right-circularlypolarized light and apply a magnetic field, along the same direction, to a gas of atoms, each absorption will force $\Delta M = +1$, whereas each emission event will have, on average, $\Delta M = 0$. Repeated absorption and reemission will

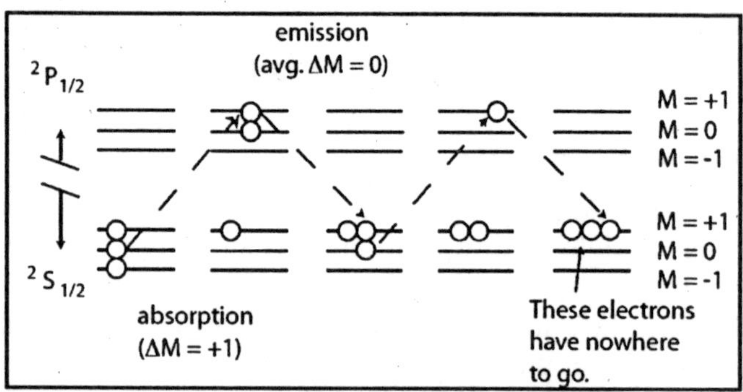

Fig. The Electrons in a Population of Atoms can be Pumped into the Highest M Level of the Ground State by Repeated Absorption and Reemission of Photons. If we only Provide Right-circularly-polarized Photons, then DM = +1 for Every Absorption.

The atoms can reemit any kind of photon, so on average $\Delta M = 0$ for reemission. A few iterations of this, and all the electrons end up in the highest M level of the ground state. This diagram illustrates optical pumping between F = 1 sublevels in Hydrogen, where I = 1/2. Rubidium works the same, except that there are more levels. "walk" the electrons into the 1s state with the highest value of M, as shown in figure 8.

Once polarized, the gas will have a total magnetization which we could, in principle, measure. However, it is much more useful to look at how well the gas absorbs the photons. As long as absorption can occur, the gas will be partially opaque to our circularly-polarized light.

Once all of the electrons are pumped into the highest M state and the gas is polarized, absorption can no longer occur. The pumped electrons have nowhere to go that would satisfy the $\Delta M = +1$ requirement. When it is in this state, the gas is transparent to our photons.

The gas being pumped is inside a cell, usually a cylinder

made of glass and sealed, along with a buffer gas. A magnetic field and circularlypolarized light are both applied along the same axis, and the transmission of the cell is measured with a photodiode. An additional radiofrequency (RF) magnetic field may be applied to drive transitions within the Zeeman levels and depolarize the gas. Our gas will remain transparent as long as the polarization is maintained. If we switch off the magnetic field, the polarization will be lost, and the gas will become opaque.

Fig. The Basic Setup for Observing Absorption Changes Due to Optical Pumping.

Similarly, pumping, and hence absorption, can occur if we scramble the electrons in the ground state by applying an RF signal that is resonant with the Zeeman splitting. In either case, the light falling on the photodetector drops when pumping is occurring, and is at a maximum when the gas is polarized.

THE QUADRATIC ZEEMAN EFFECT

Our simple model of the Zeeman splitting predicts equal splitting between the M levels of a given state. This simple model is only an approximation, valid in the limit of very weak magnetic fields.

In the weak-field limit, the Hamiltonian is diagonal in the (F, M_F) basis. However, in the strong-field limit, it is the (J, M_J, I, M_I) basis that diagonalizes the Hamiltonian, and therefore the energy level structure is different. In the transition region between these two regimes, the spacing between the levels is not equal.

Weak-Field Resonances

Start with the apparatus set up as you left it at the end of the last lab session, where you measured absorption as a function of temperature for unpolarized light.

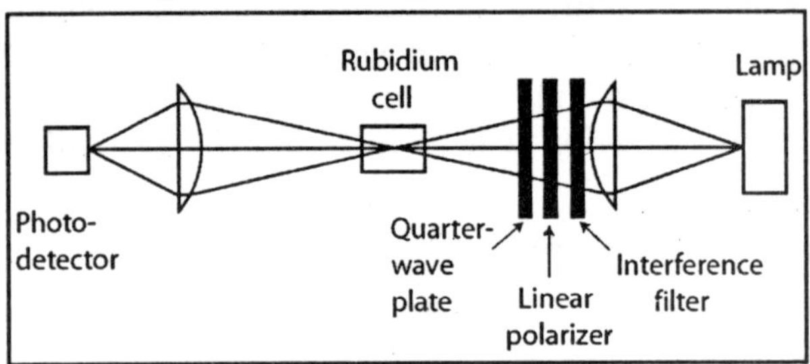

Fig. Setup for Measuring Resonant Absorption with Circularly Polarized Light.

Orient the apparatus to minimize the effect of the Earth's magnetic field. Now add the linear polarizer and quarter waveplate to produce circular polarization, making sure the orientations are correct. (Remember the description in Fowles of this arrangement, and do this "downstream" of the interference filter.) You will apply a horizontal magnetic field for this experiment, and you will need to make sure that all other magnetic fields are reduced as far as possible.

Remove any magnetic objects (anything iron or steel, along with any permanent magnets) from around the apparatus, and turn the apparatus so that its axis is North-South. You will fine-tune this alignment and use the vertical coils to compensate for the Earth's magnetic field in a little bit. Use a second linear polarizer after the quarter wave plate to check the circular polarization. If it is properly circular, the light intensity falling on the photodetector should not depend on the angle of the second polarizer. Why is this, and what would it mean if the intensity did depend on the polarizer's angle?

Use an oscilloscope to measure the output of the photodetector, and once you have verified circular polarization, don't forget to remove the second polarizer! Now null all the magnetic fields in the Rubidium cell except those along the optical axis. Set the cell temperature to 320 K, and allow it to come to equilibrium. There are two controls for the horizontal-field, one for a static field (large), and one for a slow sweep field (small). Leave the static part at zero, cover the apparatus with the black cloth, and set up a slow sweep.

Send both the sweep voltage and the output of the photodetector to an oscilloscope, and display the photodetector output versus the sweep field in the scope's XY mode. You should see a dip in the output of the photodetector when the total field (applied plus Earth's) on the cell is zero along the optical axis. This dip is the same one as the central absorption line, at B=0, in figure 10, but it will probably not be as well defined. Adjust the rotation of the apparatus and the current through the vertical compensating coils to achieve the minimum width for this absorption line, and experiment with different sweep rates to get a clean-looking line.

Take a screen shot of your best zero-field absorption line, print it out, and tape it in your lab book. This line will probably not occur at zero current, since there may be some residual Earth's magnetic field along the optical axis. Record the current, so that you can later determine this residual field. Now apply a signal to the RF coils at a frequency of about 150kHz.

Right now its amplitude is not important. You will optimize it after you have found the Zeeman resonances. Hunt around until you have found the Zeeman resonances, then adjust the RF amplitude to get the cleanest lines. Take a screenshot of the absorption curve, including the Zeeman transitions, and put it in your lab book. How does this compare with the prediction you made in the prelab?

The relative amounts of ^{85}Rb and ^{87}Rb inside the cell are not necessarily the same as those found in nature. Assuming that both isotopes have the same cross sections, what are the isotopic abundances of each species inside the cell? Measure the current in the sweep field at each Zeeman resonance, calculate the magnetic field, and compare with your expectations from the prelab. The relationship between current in the coils and magnetic field at the cell is the usual Helmholz one,

$$B = 9 \times 10^{-3} Gauss \frac{N1}{R},$$

where I is the current in amps, N = 11 is the number of turns in the coil, and R = 0.16 m is the average radius of the coils.

Do this for several RF frequencies and both isotopes. Plot the Zeeman resonance frequencies of both isotopes as a function of magnetic field. (Don't forget to take the residual Earth's field into account!) Do the slopes of these lines agree with your expectations? (Optional) Use the Zeeman resonances to calibrate both the sweep-field and static-field coils, obtaining formulas analogous to equation 1 but more accurate. You will need this if you choose to do study the intermediate-field Zeeman effect quantitatively using the Breit-Rabi equations, which will be an optional exercise later in the lab.

Intermediate-field Resonances: Quadratic Zeeman Effect

Apply a large, horizontal magnetic field using the static-field coils, and sweep around the Zeeman resonances. Does the structure of these resonances agree with your predictions from the prelab?

Do you see the expected number of lines? Do this for several different amplitudes of the RF excitation, take screenshots of your results, and comment on what you see.

Hint: You will have to determine what the best values of the static field, RF frequency and amplitude, etc. should be to do this experiment. Describe, in your lab book, why you chose the values you decided to use. (Optional) Look up, or derive, a quantitative formula for the energies of the Zeeman transitions in the intermediate-field limit for ^{87}Rb. The formulas for the Zeeman levels in an arbitrary field are known as the Breit-Rabi equations in the literature.

The Breit-Rabi equations for Hydrogen are given at the end of Chapter XII in Cohen-Tannoudji, which you read as part of your prelab exercises. Make a quantitative prediction for the spacing between the RF transitions for ^{87}Rb, and compare it with your observations. You may want to take data at more than one field value. Also recall that, in order to do this experiment properly, you need to have done the optional field-calibration exercise earlier.

RABI OSCILLATIONS

For this experiment you will want to start with an already-polarized gas, then suddenly turn on an RF field that is resonant with the Zeeman splitting and see what happens. Pick a relatively low RF frequency that will be resonant with the Zeeman splitting in the weak-field limit. Turn off the sweep field, and adjust the static field until you are on resonance for one of the Zeeman transitions. Attach a square-wave signal with an amplitude of 0 to +5 volts and a frequency of a few hertz to the modulation input on the front panel, to chop the RF signal on and off.

Look at the transmitted light intensity versus time on an oscilloscope, take a screenshot for your notebook, and explain what you see. In the prelab exercises, you made two predictions about Rabi oscillations. The first was a quantitative relationship between the frequency of the oscillations and the RF amplitude.

Chapter 17

Performance and Analysis of Thermally Tripped Mems Circuit Breakers

Electrical switches fabricated based on MEMS technologies have been reported by a number of research groups, and while many of these are intended for high frequency applications, switches suitable for DC and low frequencies have also been reported [1,2]. DC switches inevitably require direct physical contact, and a variety of configurations are possible, with specific designs choosing between lateral (in-plane) and vertical movement, and between thermal, electrostatic or other forms of actuation.

An attractive application for MEMS DC switches is in circuit protection, effectively to act as ultraminiature electro-mechanical circuit breakers. The MEMS approach provides cost, size, function and integration benefits over conventional relay-type circuit breakers. Since the maximum currents are likely to be a few amps at most, applications in electronics (e.g. on-board over-current protection) are the most suitable.

Although fuses are often used, increasingly there is a requirement for recoverable protection without component replacement . Therefore the competing technology is not the conventional fuse, but the positive temperature coefficient (PTC) device. The PTC is small, low cost, and recovers its conductive state after the over-current condition ends. However, it has serious drawbacks that a MEMS solution can overcome: slow and variable trip times, high ambient temperature dependence, and lack of controllability of re-setting.

II DEVICE STRUCTURE

The device we have developed is based on latching, laterally moving thermal actuators. The actuators are of the shape bimorph (Guckel) variety. Figure shows a schematic of the device, with the set and trip actuators, contact points, and external terminals indicated. When the actuators are latched together, the load current passes from the source connected at D to the load connected at E, either across the latch contacts or through the cold arms of both actuators via an external connection between C and B. When the load current exceeds the rated value, the trip actuator produces sufficient force to trip the latch, and the contact is broken. Resetting is achieved by applying a reset current between the set actuator terminals A and B.

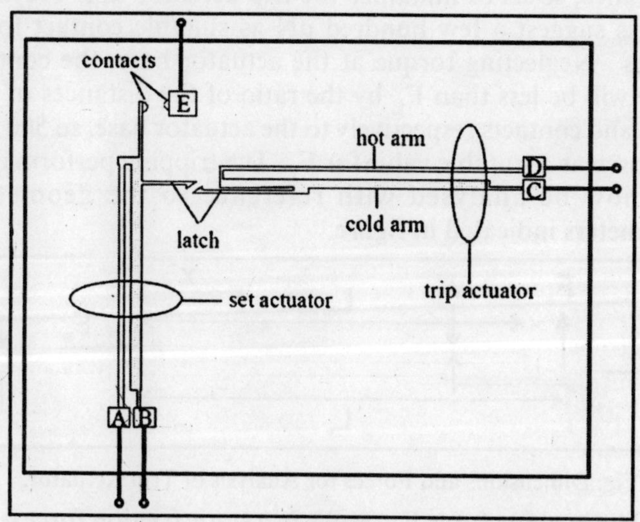

Fig. Schematic of MEMS Circuit Breaker.

The device was fabricated using nickel electroplated actuators on a silicon substrate, with gold alloy contact metallisations, with the final structure. The fabrication route is described in.

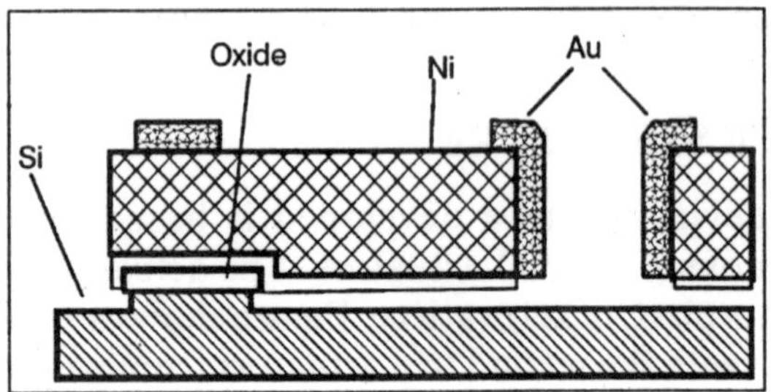

III ANALYSIS

The required trip force depends on the latching force, which can be widely varied by adjusting the stiffness of the set actuator and the displacement needed to set the latch. However, the latching force FL will determine the contact force, which in turn will set the contact resistance. The optimal F_L will be the minimum that achieves sufficiently low contact resistance, so as to minimize the trip actuator size. Previous studies suggest a few hundred μN as suitable contact force values . Neglecting torque at the actuator base, the contact force will be less than F_L by the ratio of the distances of the latch and contacts respectively to the actuator base, so 500 μN is chosen as a suitable value for F_L. The tripping performance can now be analysed with reference to the geometric parameters indicated in figure.

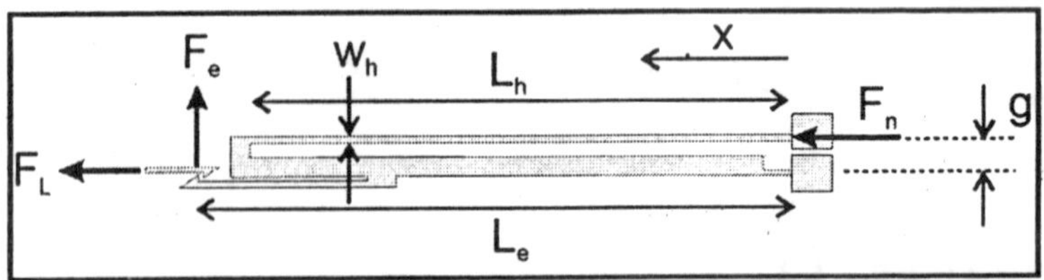

Fig. Dimensions and Forces for Analysis of Trip Actuator.

The latching force F_L results in a static friction force $\mu_s F_L$ is in turn provided by a normal force Fn which results from heating of the hot arm. Since the latch trips with minimal deflection of the trip actuators, the actuator stiffness can be neglected, so equating moments we arrive at a required $F_n = \mu_s F_L (L_e/g)$.

This is obtained from the thermal expansion of the hot arm according to:

$$F_n = E\alpha \frac{A}{L_h} \int_0^{L_h} \left(T(x) - T_A\right) dx$$

where E and α are the Young's modulus and thermal expansion coefficient of the actuator material, A is the hot arm cross-sectional area, and $T(x)-T_A$ is the temperature distribution with respect to the ambient. The hot arm expansion results from Joule heating caused by the load current I passing through the hot arm.

If we assume that the cooling is dominated by conduction down the arm to its two ends, and that these remain at ambient temperature, then the one-dimensional thermal equation is easily solved giving a quadratic variation of over temperature:

$$T(x) - T_A = \frac{I^2 \rho}{2A^2 k}\left(L_h x - x^2\right)$$

where ρ and k are the electrical resistivity and thermal conductance respectively. The integral of the over-temperature given by equation $I^2\rho L_h 2/(12A^2k)$, which we can insert in equation to derive the trip current:

$$\frac{1}{12}\frac{\alpha E}{k} g I^2 R - \mu_s F_L$$

with μ_s the coefficient of static friction, the hot arm electrical resistance $R = \rho L_h/A$, and approximating $Le \cong Lh$. For nickel the combined constant $\alpha E/k \cong 28000$ s/m^2.

For g = 20 μm, $\mu_s = 0.8$, and a target hot arm resistance of 0.1 Ω, we obtain a useful nominal trip current of Io = 0.3 A. The speed of tripping is an important performance figure. Short of a full dynamic analysis, we can obtain useful analytic approximations as follows.

As the temperature distribution approaches equilibrium, the rate of change will drop, as will the rate of increase of the actuation force. If the equilibrium value of F_e for a given current is denoted by F_{eo}, then we can approximate the rate of change of this latch release force by

$$F_e(t) = F_{eo}\left(1 - e^{-t/\tau}\right)$$

We then need to estimate the time constant τ. This we can do by approximating the rate of increase of Fe at t=0, and equating this to F_{eo}/ We then need to estimate the time constant Δ. This we can do by approximating the rate of increase of Fe at t=0, and equating this to Feo/τ.

At the initial stages of heating, the temperature rise will be relatively uniform along the hot arm. Then the integral is given simply by $(T(x)-T_A)L_h$. Since the conductive cooling is initially negligible, the rise in excess heat per unit volume can be equated to the electrical power according to:

$$\rho_m c_p \frac{dT}{dt} = \frac{I^2 R}{A^2}$$

where ρm and cp are the mass density and heat capacity respectively. We can then derive dFe(t)/dt from dT/dt, and thus obtain our approximation to t as:

$$\tau = \frac{1}{12} \frac{\rho_m c_p}{k} L_h^2$$

If the required trip force is FeT, the trip time can be derived from as τ-ln(F_{eo}/(F_{eo}-F_{eT})). Since the equilibrium force is proportional to I^2, we obtain:

$$t = \tau \ln \frac{I^2}{I^2 - I_o^2}$$

Taking values for ρm, cp and k of 8900 kg/m^3, 0.45 kJ/kgE"K and 900 W/mE"K respectively, and L_h = 2 mm, the time constant Δ = 18 ms. For large overcurrent rations I/Io, reduces to t = τ(Io/I)^2. The devices were successfully fabricated with dimensions as described above. Electrical tests gave contact resistances below 1 Ω in nearly 50% of working devices, and trip currents near the design value at 300 – 400 mA. Figure shows a prototype device in the set (latched) state.

Trip times were measure as a function of the ratio of applied current I to minimum trip current Io. The results are shown in figure, along with the theoretical prediction of equation. The fit is close for lower currents, but for over-current ratios above 2, the trip time is longer than predicted and appears to level off. One possible cause is the time taken for the mechanism to displace the small distance of the latch mating surfaces (20 μm).

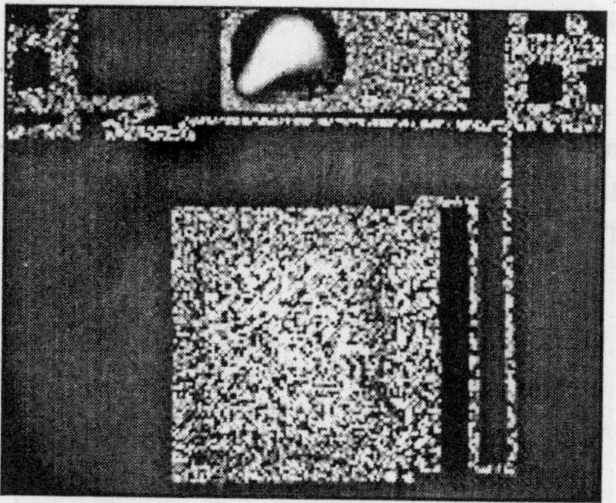

Fig. Optical Micrograph of Circuit Breaker Latch Area and Contact in Closed State.

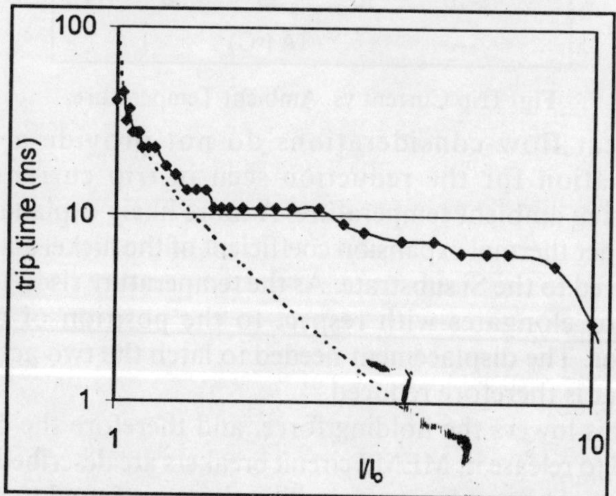

Fig. Trip time vs. over-Current ratio: Experimental Values (solid line) and model (Dashed Line).

However, this time can be shown to be insignificant if inertial factors dominate. The total mass of the actuators is very low (≈ 0.25 mg), and the driving force will be a significant fraction of Feo, since the dynamic coefficient of friction is significantly less than the static one.

Taking the driving force as 100 μN gives a time to release the latch of ≈ 40 μs. A noninertial "sticking" effect in the latch may instead be the cause. In any case the measured trip times are orders of magnitude less than for PTCs of similar rating.

The variation of trip current with ambient temperature was also measured, and the results are given in figure. The slope is about 10X less than for PTC devices at 0.15 %/K.

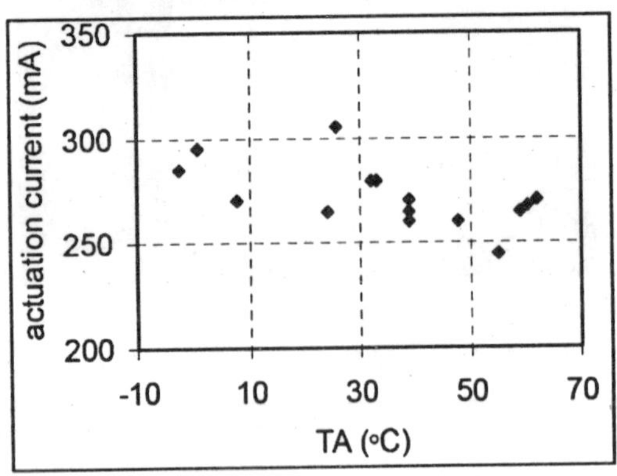

Fig. Trip Current vs. Ambient Temperature.

Heat flow considerations do not provide a ready explanation for the reduction seen in trip current with increasing ambient temperature. A more likely explanation is the higher thermal expansion coefficient of the nickel actuators compared to the Si substrate. As the temperature rises, the trip actuator elongates with respect to the position of the set actuator. The displacement needed to latch the two actuators together is therefore reduced.

This lowers the holding force, and therefore the current needed to release it. MEMS circuit breakers are described based on thermal bimorph actuators. Trip times are found to be low, and to have a low temperature dependence, compared to PTC devices. A model for trip time is presented which gives good correspondence to measurements for low over-current ratios, but trip times are found to drop less quickly than predicted for high over-currents.

Index

A

B

C

D

E

F

G